Weapons for Peace,
Weapons for War

Weapons for Peace, Weapons for War

The Effect of Arms Transfers on War Outbreak, Involvement, and Outcomes

Cassady Craft

Routledge
New York, London

Published in 1999 by
Routledge
29 West 35th Street
New York, NY 10001

Published in Great Britain by
Routledge
11 New Fetter Lane
London EC4P 4EE

Printed in the United States of America on acid-free paper.

Library of Congress Cataloging-in-Publication Data

Craft, Cassady.
    Weapons for peace, weapons for war : the effect of arms transfers on war outbreak, involvement, and outcomes / by Cassady Craft.
            p.   cm.
    Includes bibliographical references and index.
    ISBN 0-415-92258-5 (cloth). — ISBN 0-415-92259-3 (pbk.)
    1. Arms transfers. 2. War. 3. Weapons industry. I. Title.
UA10.C73 1999
355.02'7—dc21   98-49428
                    CIP

Contents

Acknowledgments

Popular culture has the vision of the scholar as someone who is sequestered in the ivory tower, engaged in a more or less solitary search for the truth. Nothing could be further from reality. Scholars have lives (their students don't believe this, though), friends, family, and of course, colleagues. As research and writing is of necessity a recursive process, there are innumerable people who have made an impact on this work and yet do not bear any responsibility for its shortcomings. I would like to thank the following for their encouragement, advice and assistance in the completion of this research: Robert Grafstein, Vally Koubi, Jerry Legge, Bill Chittick and Han Park. The Director of the Center for International Trade and Security, Gary Bertsch, and Associate Director Richard Cupitt were kind enough to provide facilities, travel support, friendship and advice without which this project would have languished. Gary Bertsch, in particular, made my years at the University of Georgia productive, rewarding, and in the end, worthwhile. He and Joni were kind enough to extend their hospitality, graciousness and friendship. Other Center colleagues were kind enough to sit through brown-bag presentations, strategize about the dissertation process, and to make it easy to work around their busy schedules by allowing an occasional encroachment. Of these colleagues, Keith Wolfe and Jonathan Benjamin Alvarado stand out as especially shining examples of the true nature of a scholar—adept in research, teaching and service. Keep setting the bar higher, guys.

Ian Anthony, Pieter and Siemon Wezeman of the Stockholm International Peace Research Institute were hosts during my brief visit to Sweden, and saved this project when the "data" that I thought I'd come away with was not what I thought. For this, I owe a great debt. Jonathan Gill and Christopher McDonald provided intellectual sustenance by being around to talk, often expressing interest in the mundane details of data and theory, and showing how weapons sales really do matter. Their companionship in demonstrating the capabilities of military weaponry as old as our grandfathers continues to be revealing of both the destructive power of weapons themselves and the rejuvenating power of friendship. In this respect, I can only hope that I gave as good as I got and do everything in my power to make sure that our friendship is lifelong. Robert Harkavy and Frederic Pearson proved to be intellectual pathfinders and were kind enough to read and comment on early drafts.

I envy their students. Ron Page of the Correlates of War Project provided data updates and answered innumerable questions. Numerous "lurkers" on the Scientific Study of International Processes internet discussion group also were kind enough to share insights through the electronic medium. My students at the Sam Nunn School of International Affairs at the Georgia Institute of Technology seemed willing to forgive as I thought aloud concerning the intricacies of arms transfers data, statistical significance, and the substantive importance of the study of conflict processes.

Dr. and Mrs. Gerry Craft taught their youngest boy, among a myriad of other important things, to finish what he started. Anyone who has been a graduate student knows that sometimes that is the key in completing an advanced degree. Through the years, Russell Pierson has shown me how a true cowboy does his work, and Bernice, my grandmother, exhibited all of the qualities of a wonderful, beautiful lady. They are also responsible for my early exposure to international affairs by virtue of their allowing a twelve year old to travel along with them around the Pacific rim.

Last, but most importantly, Suzette Grillot provided coaching and encouragement as wife, confidant, and friend. Her many trips to the mound to share advice and give me a breather allowed me to avoid unnecessary pitfalls and, eventually, to notch up a complete game. Her previous experience in writing a dissertation inspired me because she bore our child during the process. Our daughter, Hannah, also had indelible impact on this project by rearranging notes and drafts, pounding on keyboards and otherwise filling my life with happiness, fun, and play. Thanks, little girl, for never allowing me to forget the important things in life.

1

Introduction: Weapons for Peace, Weapons for War

Every modern war threatens to involve half the world, bring disaster to world economy, and blot out civilization. The question is urgent then: What will be done about the armaments industry?

H.C. Engelbrecht and F.C. Hanighen, *Merchants of Death*[1]

[The] relative importance [of the abuses of private arms manufacturers] in the causation of modern war has probably been greatly exaggerated .... While private arms-trading ... [has] led to abuses, it seems probable that on the whole [it] has tended to stabilize the balance of power rather than to disturb it by equalizing the defensibility of states. Control of these activities by national governments would tend to increase international tensions.

Quincy Wright, *A Study of War*[2]

The Argument

Why study the effect of arms transfers[3] on war? As noted by Engelbrecht and Hanighen more than half a century ago, trends in the global arms trade are carefully watched because of expectations concerning the outbreak and outcomes of war.[4] Most analysts conclude that increasing global arms sales enhances military capabilities around the world, thus leading to an increase in the likelihood of war.[5] War, seen as a public phenomenon, produces negative externalities, such as demographic dislocations, economic distortions, contamination from the use of weapons of mass destruction, and potentially the fundamental reordering of the international system.[6] If arms transfers heighten the likelihood of war outbreak or intensify these, or other, consequences of war, then their study should play an integral role in peace research. Finally, because war has at least since the days of Clausewitz been conceived as "politics by other means," it must be considered as a policy option by statesmen.[7] If arms transfers alter leaders' decision-making calculus in assessing probable costs and benefits of militaristic policies (as asserted by Stockholm International Peace Research Institute), then researchers should investigate the precise relationships.[8]

At the same time, Wright's argument makes a powerful counter-claim—one backed by important empirical evidence.[9] Given the certainty with which those who espouse the "merchants of death" argument make their case, one would expect clear, consistent, and systematic empirical evidence to this effect. Unfortunately, this is not the case. The evidence that does exist instead tends to be based on anecdotes; or when more rigorous, has severe shortcomings in temporal inclusiveness, theoretical confusion, and weak and mixed results. Furthermore, Wright's injunction against government intervention into the international arms trade is based on some of the most widely held beliefs in the study of history: that states have the right to arm for defensive purposes so that they may guarantee the security of their constituents in an anarchical international system, that strategies of deterrence supported by military power enhance national security, and that governments should refrain from needlessly imposing on the economic welfare of their citizens.

Given the above reasons, and undoubtedly many more, studying the effects of arms transfers on war involvement and outcomes is as important today as it was 65 years ago when Engelbrecht and Hanighen penned *Merchants of Death*. Ultimately, the extreme risks of modern war and the institutionalized power and independence of what is increasingly seen as a global military industrial complex contribute to the question's immediacy.[10]

Contending Viewpoints

While scholars have previously studied a number of questions concerning arms transfer effects on war involvement and outcome, none has systematically surveyed these relationships in a rigorous, systematic, and empirical manner. Scholars posit many theoretical relationships between arms transfers and war. The most prominent assertion based on anecdotal evidence is that there are indeed observable relationships between the two phenomena. According to one major work on the subject, the weapons trade is associated with increased likelihood of war participation and also an increased severity and magnitude of resulting wars.[11] If so, then knowing the precise strength and orientation of the relationship between arms transfers and war outbreak and outcomes is beneficial so that we may mitigate their effects.

Other researchers make contradictory arguments.[12] There is a considerable literature concerning theories of deterrence that build on the Roman motto, "If you desire peace, prepare for war." According to these theorists, states arm so that others fear the costs of attacking them and therefore attain a peaceful, if wary, existence. Consequently, if peace stems from the sale of weapons, then better understanding of the precise correlation should allow beneficial policies to be enacted.

If there is no demonstrable relationship between the arms trade and war, then actions to limit the sales of the weapons industries of the world harm the public interest. Policies to control the weapons trade must in any case be carefully crafted because restraints on weapons sales undoubtedly have economic consequences, such as unemployment, underperformance of important sectors of the economy, and lower standards of living wrought by forgone sales and profits.[13] Few governments can afford to place such penalties on their constituencies without firm evidence to their effect. Why might there be no effect even if weapons are seen as necessary conditions for war or peace? First, there is the inhibition against employing resources, such as weapons, in a high-risk environment typified by war. Some leaders, perhaps, choose to "lose" the war, or negotiate peace, rather than to risk losing the weapons that they have so carefully hoarded. Saddam Hussein's control of the materiel and personnel of his elite Republican Guard units in the Persian Gulf War provides a prominent example. Second, some argue that Third World states lack the technological skill necessary to effectively employ under wartime conditions many modern weapon systems that they acquire.[14] The purpose of these weapons, instead, is to give prestige to the ruling regime or to intimidate either international or domestic foes.

Regardless of the reasoning behind their actions, or lack thereof, it is clear that the states that supply the majority of the international trade in weapons, even though part of a 50-year-old arms *control* regime, find it difficult to either control the "arms merchants" or to convince them that they should refrain from making sales. As William Hartung has noted in *And Weapons for All*, those who are interested in these questions have all too often seen circumstances similar to those that occurred after the conclusion of the Persian Gulf War (Operation Desert Storm) where the U.S. government simultaneously sought to create an arms transfer control regime among major conventional weapons suppliers while assigning State Department, military, and other executive branch officials to help *promote* the sale of U.S.-manufactured weapons in countries such as Saudi Arabia and Taiwan.[15]

The Analytical Controversy

As in many areas of study in the social sciences, there are controversies among scholars of arms transfers and war concerning what types of analysis are most appropriate for various research questions. According to some, weaknesses of the arms transfers data preclude rigorous statistical exercises.[16] Consequently, some scholars rely on historically oriented analysis.[17] A second group supplements historical case studies with various types of descriptive statistics in order to show the relative magnitudes of arms transfers or rough correlation of military forces.[18] Finally, others claim that such analysis is confined by too many variables and too few cases; i.e., that because of the plethora of variables under consideration for each case under examination, little or no generalization can result from descriptively oriented case studies.[19] As a result, a number of scholars have performed statistical analysis of relationships of arms transfers and various conflict processes, although at times in a haphazard manner.[20]

The analysis performed in this work adheres to the latter philosophy of empirical examination for several reasons. First, while acknowledging the desirability and utility of theory building based on comparative research, there is an undeniable need for general empirical analysis in gaining basic insights into general phenomena. Because comparative analysis, as noted by Brzoska and Pearson, "only makes sense when it is about a certain topic under certain circumstances," a more complete understanding of the relationship(s) between variables requires examination in a more general context.[21]

The second reason is, quite simply, that there are many assertions, but serious gaps in the research concerning the systematic impact of arms transfers on war outbreaks and outcomes. As noted by Kinsella, this research has to date been neither cumulative nor consistent.[22] It is hoped that by performing a more extensive statistical survey of these relationships, this book will fill gaps in our general knowledge. In terms of war outcomes, this is virtually assured, given the absence of empirical research on related questions.

Finally, the subject of the impact of weapons exports or imports on war is too pressing to relegate to the back burner of scientific research until we have better data, especially if we must rely on such institutions as the UN register of conventional weapons sales for the acquisition of that data. Initially, our analysis would be delayed by decades while waiting for sufficient numbers of cases to examine with statistical rigor. In addition, there are serious doubts as to whether the UN register, which relies on voluntary reports of transfers by supplier and recipient states, will *ever* supplant in quality existing sources such as those provided by other institutions like SIPRI (Stockholm International Peace Research Institute) and the U.S. Arms Control and Disarmament Agency (ACDA).[23]

A second major controversy revolves around a deceptively simple question: What is war? The most commonly accepted definition of war is that attributed to Clausewitz as state-centric "continuation of politics by other means."[24] However, military historians have pointed out that (a) this is a subtle mistranslation of Clausewitz; and (b) that what the great German thinker was proposing was a definition of what war *should be* in the Western world of the early nineteenth century rather than what it actually was, had previously been throughout history, and is today.[25] War is actually, according to these scholars, much closer to other definitions that encompass a more fundamental aspect of human relations prior to the state—as a deadly struggle for power, and all that comes with it, within society. As such, and because war predates the state, to confine its definition to "a violent conflict between states," as do Ropp and others, is a distortion of what war actually is in preference to what many would have it be.[26] This study, as will become obvious, declines to accept the state-centric definition of war and its logical distinction between interstate "war" and civil "conflicts" or "violence."[27] It instead relies on a broader view of war that encompasses both inter- and intrastate war as violent struggles for political power within societies.

Adopting this broader definition of war is important for several reasons. First, as noted above, the analytical distinction between interstate and civil war is, at best, tenuous. Instead, many wars take on the attributes of both. Lebanon presents a case where the civil war and international war elements of the conflict were so completely inter-mingled that it is almost impossible to separate the two. The Vietnamese situation was almost as complex, as are the wars in the former Yugo-slavia and former Soviet Union (Nagorno-Karabakh, Tajikistan, Georgia, Chechnya, etc.). This class of hybrid complex wars (civil wars that explode internationally, international wars that implode domestically, and wars that take place at both levels simultaneously) have been prevalent during the 1950–1992 period.

There is a second, normative viewpoint that indicates that interstate and civil war should not be separated (nor denigrated via reversion to some other label such as "militarized violence"), because to do so indicates that there is a difference between a state using imported weapons to kill citizens of another state and using them to kill one's own population. While some policymakers are able to avoid this question altogether by asserting that arms sales should be considered only on their economic merit,[28] others—especially in the democratic states that supply most of the world's weapons—find it difficult to avoid the questions posed by constituents concerned about the domestic and foreign policy aspects of selling weapons abroad. These policymakers must be, therefore, concerned with the predicted political effects of weapons sales on supplier and recipient countries in terms of the political backlash that policies are likely to cause. When imported weapons are used to *either* engage in cross-border disputes *or* to repress dissent within the recipient country, supplier state policymakers are certain to sow a whirlwind of public reprobation. Thus, it is important for them to assess the relationship between weapons and war at both the domestic and interstate levels.

What Follows

Given the divergent assertions and findings of prior research, a new and more comprehensive look at the effect of arms transfers on war is needed. The following chapters provide a serious attempt to do this by surveying the relationship in three ways: global relationships, supplier and recipient dynamics, and dyadic interactions. Reliance on evidence at multiple levels of analysis provides an important means to cross-check findings concerning the types, direction, and strength of variance in the relationship between arms transfers and war involvement and outcomes.

At the global level, it is important to know what general trends exist between arms transfers and war so that we may determine whether some basic premises of international relations scholarship hold. While there is no true systemic theory to apply to this relationship, certain subsystemic theories inform us to what the true correlations may be. Some argue that levels of arms transfers rise prior to the outbreak of wars because states foresee conflict and begin arming so that they may participate more advantageously. Others insist that increased weaponization provides for deterrence and, therefore, peace. There are also those that observe that arms transfers often occur *after* wars begin because states require large supplies of weapons to fight. If this is the primary relationship between arms transfers and war, one may conclude that the former has little or no causal effect on the latter. Finally, some scholars argue that increases in global arms transfers should lead to increased bloodiness, severity, and magnitude of wars as the influx of weapons serves to "add fuel to the fire" of ongoing conflagrations.[29]

Chapter 2 investigates the veracity of the various claims made by theorists to determine whether there are observable patterns between arms sales and wars at the global level. In examining this relationship at this level and using aggregate arms sale amounts and global war indicators, the most general findings are produced. Using correlational methods appropriate for measuring the extent and direction of covariation, it is found that increases in global arms transfer levels are mildly associated with increased amounts of war around the world. However, as global arms transfer patterns also are shown to *follow* increases in the amount of war around the globe, causal statements cannot be made. Considering war outcomes, the findings in Chapter 2 indicate that increases in global arms transfers are not associated with bloodier wars, but are positively correlated with greater amounts and magnitude of war.

The third chapter provides an analysis of the means by which arms transfers influence the foreign policies of supplier and recipient states. Do arms sales make suppliers more likely to become militarily involved in wars of recipients, in a manner reminiscent of the American entrance into the war in Vietnam? Are arms transfers more likely to occur after war termination than in other years, as many "resupply models" suggest? Do imports of weapons precede war involvement or make recipients more apt to participate in wars? Concerning war outcomes, do increased weapons imports prior to the outbreak of wars make them longer or bloodier? Are antebellum (prewar) arms sales typically influential in determining the winners of wars? Questions such as these, and empirical evidence as to what the answers are, provide an important venue for foreign policy decision making.

Chapter 3 provides empirical evidence that debunks the credibility of many of the competing claims. In fact, the majority of the evidence presented in that chapter indicates that there is really little to support these claims. Arms transfers are not more likely to follow war termination, nor are suppliers more likely to participate in wars during years of increased weapons sales. Weak evidence exists that indicates increased arms imports do signal that recipients are more likely to participate in war, but these findings require additional research before we can be assured of the precision of this relationship. When it comes to war outcomes, there seems to be little, if any, relationship between arms transfers prior to the war outbreak and such phenomena as the length, bloodiness, or winners of wars. Indeed, the only relationship found to exist between these issues in Chapter 3 insinuates that countries that do not increase their weapons imports in the three- to five-year period immediately preceding a war are more likely to emerge victorious in a conflict.

The analysis in Chapter 4 concerns the relationship between arms transfers and war involvement and outcomes. It attempts to determine whether arms transfers affect the antebellum perception of "military balance" between states, whether arms transfers affect the likelihood of war involvement directly, and whether arms transfers *after* the initiation of the war influence the outcomes of the war. There have previously been many claims about the effects that arms transfers have on the "military balance" between states, but very few scholars have actually explored the particular dynamics. Chapter 4 does this while keeping in mind that change of "military balance" is little more than assessments of what the likely war outcomes are, given correlations of opposing forces at two points in time: before an event, such as the import of weapons by one side; and after. According to these theories, when one leader perceives an opportunity to defeat another because of an advantage gained by the import of weapons, he or she is more likely to initiate military operations against that opponent. Yet all of this theory presumes alterations in the military balance are (1) perceptual and measurable, and (2) distinguishable and discernible. Because we find that arms transfers have no distinguishable effect on the measurement of the ex ante military balance, we reject the proposition that alterations in the military balance intervene between arms transfers and war involvement.

Next, we examine the direct relationship between arms transfers and war involvement based on SIPRI's theory, which states that arms transfers heighten leaders' awareness of military options to foreign policy problems, therefore making them more likely to use force when confronted with disputes involving their neighbors.[30] We find evidence to indicate that the SIPRI theory may indeed retain its plausibility. Finally, we investigate the proposition that arms transfers imported after the beginning of a war cause wars to last longer and to become more bloody. In Chapter 4, the findings indicate that there is little, if any, evidence for these assertions. Neither do arms transfers appear to be related to shorter or less bloody wars. In short, there seems to be no relationship between the arms trade and these types of war outcomes.

Chapter 5 assesses the overall impact of the previous findings by examining the role of differing weapons types in *future* warfare. Based on the forecasts of prominent military scientists, defense analysts, historians, policymakers, and political scientists, a typology of future wars is devised. We examine efforts to control the trade in weapons in light of the future war typology and make recommendations for the future of arms control based on this research.

This work provides a renewed effort to examine the question of whether arms transfers affect war involvement and outcomes in a comprehensive manner. As with any attempt of this sort, it necessarily falls short of answering the innumerable questions that arise along the way. It is perhaps a matter of revealed truth that investigations, such as those offered here and by predecessors such as Brzoska and Pearson, Harkavy, and Laurance, of questions concerning the relationship between arms transfers and war should continue in the effort to provide conclusive results, for several important reasons.[31] First, if the positive association between arms transfers and war involvement found here holds up to the rigors of further empirical research, then peace researchers should encourage governments to adopt policies that restrain weapons sales. If future research reverses these findings and establishes conclusively that the true relationship is robust and negative, then policymakers should be encouraged to promote peace through the creation of deterrent dynamics by promoting the sale of weapons. Finally, if the rigors of additional future research indicate that there is no causal relationship between arms transfers and war involvement, then scientists should encourage government restraint in creating policies that would moderate the manufacture, marketing, and sale of weapons. For if the global weapons industry is benign (in the sense that weapons do not contribute to the onset or worsening of the effects of conflict), then government politico-economic policies to check such sales would be to the detriment of their constituencies via forgone contributions to GNP, income, balance of trade, and employment.

This research indicates that the truth is very hard to discern. It appears, overall, that arms transfers *are* associated with war involvement, but not the worsening of war outcomes. The reader may disagree with the approaches taken here; the use of models, simulations, military science, and statistical methods; and with the overall results. However, if this work serves to encourage discussion of these important issues, then it has achieved a most important goal.

Engelbrecht and Hanighen observed in 1934 that the dangers of war were enhanced by the rapid, thorough, and largely unregulated transformation of war-making capacity in the global system. Largely wrought by the weapons makers and often with the tacit support of governments, the "death machines" of 1934 represented the "acme of scientific achievement."[32] How much more so today, as private and national laboratories, bureaucratic strategists, and university researchers pitch in to systematize and accelerate the development of ever more effective, and usually more deadly, weapons? For most, it is their earnest desire that these weapons will promote peace; the greatest fear of many is that they incite war. This book is written in the hope that the uses of scientific research can contribute to the adoption of enlightened policies that will help to avoid the carnage, horrors, and destruction of war. We must remember that war has another definition, as well, one rendered by William T. Sherman as he ravaged the southern portion of his own country during the American Civil War: "War is hell."

Notes

1. H.C. Engelbrecht and F.C. Hanighen, *Merchants of Death: A Study of the International Armament Industry* (New York: Dodd, Mead, 1934).

2. Quincy Wright, *A Study of War*. 2 vol. (Chicago: University of Chicago Press, 1942).

3. Throughout this study, I use the terms *arms sales, arms trade, arms transfers,* and *weapons trade* interchangeably to denote the trade in conventional weapons, including ballistic missiles. Weapons of mass destruction transfers are not included by these terms.

4. Engelbrecht and Hanighen, *Merchants of Death*, 1934.

5. For example, see Frederic Pearson, Michael Brzoska, and Christopher Crantz, "The Effect of Arms Transfers on Wars and Peace Negotiations," in Stockholm International Peace Research Institute (hereafter SIPRI), *SIPRI Yearbook 1992, Armaments and Disarmament* (Oxford: Oxford University Press, 1992); Michael Brzoska and Frederic Pearson, *Arms and Warfare: Escalation, De-escalation, and Negotiations* (Columbia: University of South Carolina Press, 1994); Christian Catrina, "Main Directions of Research in the Arms Trade," *Annals of the American Academy of Political and Social Science* 535 (1994): 190–205.

6. William Thompson, "The Consequences of War," *International Interactions* 19 (1993): 125–47.

7. Bruce Bueno de Mesquita, *The War Trap* (New Haven, CT: Yale University Press, 1981); Bruce Bueno de Mesquita and David Lalman, *War and Reason* (New Haven, CT: Yale University Press, 1992).

8. SIPRI, *The Arms Trade with the Third World* (Stockholm: Almquist and Wiksell, 1971).

9. Wright, *A Study of War*, 1942.

10. Engelbrecht and Hanighen, *Merchants of Death,* 1934; Richard Bitzinger, "The Globalization of the Arms Industry: The Next Proliferation Challenge," *International Security* 19 (1994): 170–98.

11. Brzoska and Pearson, *Arms and Warfare*, 1994; see also Pearson, Brzoska, and Crantz, "The Effect of Arms Transfers," 1992.

12. See E.H. Carr, *The Twenty Years Crisis, 1919–1939: An Introduction to the Study of International Relations* (London: Macmillan, 1946); James Foster, "New Conventional Weapons Technologies: Implications for the Third World," in *Arms Transfers to the Third World: The Military Buildup in Less Industrial Countries*, ed. Uri Ra'anan, Robert Pfaltzgraff, and Geoffrey Kemp (Boulder, CO: Westview Press, 1978), 65–84; Hans Morgenthau, *Politics Among Nations: The Struggle for Power and Peace* (New York: Knopf, 1948); Wright, *A Study of War*, 1942; and Kenneth Waltz, *Man, the State, and War: A Theoretical Analysis* (New York: Columbia University Press, 1959).

13. Christian Catrina, *Arms Transfers and Dependence* (New York: United Nations for Disarmament Research, 1988).

14. Foster, "New Conventional Weapons Technologies," 1978, 65–84.

15. William Hartung, *And Weapons for All: How America's Multibillion-Dollar Arms Trade Warps Our Foreign Policy and Subverts Democracy at Home* (New York: HarperCollins, 1994).

16. Brzoska and Pearson, *Arms and Warfare*, 1994; Frank Blackaby and Thomas Ohlson, "Military Expenditure and the Arms Trade: Problems of Data," *Bulletin of Peace Proposals* 13 (1982): 291–308. For those who maintain this point, what follows is, of course, little more than a model-building exercise. However, when and if "better" data become available, the models presented here should increase in their usefulness even to the skeptics of statistical analysis.

17. See, for example, Brzoska and Pearson, *Arms and Warfare*, 1994; Pearson, Brzoska, and Crantz, "The Effect of Arms Transfers," 1992; Hartung, *And Weapons for All* (1994); Keith Krause, "Military Statecraft: Power and Influence in Soviet and American Arms Transfer Relationships," *International Studies Quarterly* 35 (1991): 313–36; William Quandt, "Influence Through Arms Supply: The US Experience in the Middle East," in Ra'anan, Pfaltzgraff, and Kemp, eds., *Arms Transfers to the Third World*, 1978, 121–30; Uri Ra'anan, "Soviet Arms Transfers and the Problem of Political Leverage," in *Arms Transfers to the Third World*, ed. Ra'anan, Pfaltzgraff, and Kemp, 1978, 131–56; and Miles Wolpin, *America Insecure: Arms Transfers, Global Interventionism, and the Erosion of National Security* (London: McFarland and Company, 1991).

18. Examples are: Robert Harkavy, *The Arms Trade and International Systems* (Cambridge, MA: Ballinger Publishing Company, 1975); Robert Harkavy, "Recent Wars in the Arc of Crisis: Lessons for Defense Planners," in *Defense Planning in Less-Industrialized States: The Middle East and South Asia*, ed. Stephanie Neuman (Lexington, MA: D.C. Heath, 1984), 275–300; Edward Laurance, *The International Arms Trade* (New York: Lexington Books, 1992); Stephanie Neuman, *Military Assistance in Recent Wars* (New York: Praeger, 1986); Stephanie Neuman, "Third World Military Industries: Capabilities and Constraints in Recent Wars," in *The Lessons of Recent Wars in the Third World, Volume II*, Robert Harkavy and Stephanie Neuman, eds. (Lexington, MA: Lexington Books, 1987); Stephanie Neuman, "The Role of Military Assistance in Recent Wars," in *The Lessons of Recent Wars, Volume II*, ed. Harkavy and Neuman 1987; and Stephanie Neuman, "Arms, Aid and the Superpowers," *Foreign Affairs* 66 (1988): 1044–66.

19. David Kinsella, "Conflict in Context: Arms Transfers and Third World Rivalries During the Cold War," *American Journal of Political Science* 38 (1994): 557–81; William Baugh and Michael Squires, "Arms Transfers and the Onset of War Part I: Scalogram Analysis of Transfer Patterns," *International Interactions* 10 (1983): 39–63.

20. See Baugh and Squires, "Arms Transfers and the Onset of War Part I," 1983; William Baugh and Michael Squires, "Arms Transfers and the Onset of War Part II: Wars in Third World States, 1950–65," *International Interactions* 10 (1983): 129–41; Kinsella, "Conflict in Context," 1994; David Kinsella and Herbert Tillema,

"Arms and Aggression in the Middle East," *Journal of Conflict Resolution* 39 (1995): 306–29; Ronald Sherwin, "Controlling Instability and Conflict Through Arms Transfers: Testing a Policy Assumption," *International Interactions* 10 (1983): 65–99; Ronald Sherwin and Edward Laurance, "Arms Transfers and Military Capability: Measuring and Evaluating Conventional Arms Transfers," *International Studies Quarterly* 23 (1979): 360–89; Philip Schrodt, "Arms Transfers and International Behavior in the Arabian Sea Area," *International Interactions* 10 (1983): 101–27.

21. Brzoska and Pearson, *Arms and Warfare*, 1994.

22. Kinsella, "Conflict in Context," 1994.

23. The data currently provided by the UN arms register suffer from the fact that many states do not submit transfer records for foreign policy reasons. The United Nations has proven historically relatively powerless in the face of such obstacles due to sovereignty issues. Moreover, because of the varying levels of aggregation of the data provided by the states in their voluntary submission, these data are inconsistent and therefore less useful than they could be. Finally, while hardly unanimous, there is some agreement as to what types of data concerning arms trade are needed. Michael Brzoska, "Arms Transfer Data Sources," *Journal of Conflict Resolution* 26 (1982): 77–108; Edward Fei, "Understanding Arms Transfers and Military Expenditures: Data Problems," in *Arms Transfers in the Modern World*, ed. Stephanie Neuman and Robert Harkavy (New York: Praeger, 1979), 37–48; and Edward Laurance and Ronald Sherwin, "Understanding Arms Transfers Through Data Analysis," in *Arms Transfers to the Third World*, ed. Ra'anan, Pfaltzgraff, and Kemp, 1978, 87–105, provide important guides to thinking in this area, and the UN register is clearly *not* an improvement over current sources, such as the various SIPRI databases in areas such as provision of data on individual transfers, precise date of delivery, etc.

24. Carl von Clausewitz, *Vom Krieg* [On War], trans. Michael Howard and P. Paret (Princeton, NJ: Princeton University Press, 1976).

25. See especially Martin van Creveld, *On Future War* (London: Brassey's, 1991); and John Keegan, *A History of Warfare* (New York: Vintage Books, 1993).

26. Theodore Ropp, *War in the Modern World* (New York: Collier Books, 1964); see also J. David Singer, ed., *Correlates of War I: Research Origins and Rationale* (New York: Free Press, 1979); J. David Singer, ed., *Correlates of War II: Testing Some Realpolitik Models* (New York: Free Press, 1979); and J. David Singer, ed., *Explaining War: Selected Papers from the Correlates of War Project* (Beverly Hills: Sage Publications, 1979).

27. For similar views, see Lewis Richardson, *Arms and Insecurity* (Pittsburgh: Boxwood, 1960); Lewis Richardson, *Statistics of Deadly Quarrels* (Chicago: University of Chicago Press, 1960); Harvey Starr, "Revolution and War: Rethinking the Linkage between Internal and External Conflict," *Political Research Quarterly* 47 (1994): 481–507; Harvey Starr and Benjamin Most, "Diffusion, Reinforcement, Geopolitics, and the Spread of War," *American Political Science Review* 74 (1980): 609–36.

28. For politicians such as these—or like-minded analysts—the normative aspects of the question posed here matters very little, if at all. However, even if one is to take their view, the findings of this research deserve merit due to their original contribution to the scientific literature concerning the question of the effect of arms transfers and war. Of course, it would be possible to make slight alterations of the models presented here to independently assess the impact of arms transfers on interstate or civil wars. This author, however, believes that to do so would introduce to the analysis a myriad of distortions that would make the findings irrelevant to policymakers and citizens—those who should be the ultimate consumers of political science research.

29. As noted by Thompson, "The Consequences of War," 1993, "the lion's share of social science attention has been and continues to be oriented toward explaining the causes of war" at the expense of the study of war's consequences. Likewise, most studies of the impact of arms transfers on war have considered implications for war outbreak and neglected questions of war outcomes. The research presented here provides insight into both aspects of arms transfer effects.

30. SIPRI, *The Arms Trade with the Third World*, 1971.

31. Brzoska and Pearson, *Arms and Warfare*, 1994; Harkavy, *The Arms Trade and International Systems*, 1975; and Laurance, *The International Arms Trade*, 1992.

32. Engelbrecht and Hanighen, *Merchants of Death*, 1934, 257.

2

The Effects of Arms Transfers
on War Outbreak and Outcomes
in the International System, 1950–1992

When international relations become strained, the business of the arms industry generally improves. Hence when nations manage to get along with a minimum of friction, the arms makers have sometimes not hesitated to stir up trouble. Every nation has "natural" or "hereditary" enemies. The arms makers need do no more than to point out the increasing armaments of the "enemy" and the virtual "helplessness" of the "threatened" country and before long there is vigorous action for "preparedness," which in turn means business for the arms merchants. Sometimes it is sufficient to sell to the "enemy" the latest engines of war and then to apprise the other government of that fact.

H.C. Engelbrecht and F.C. Hanighen, *Merchants of Death*[1]

There is now ... a growing appreciation that the arms trade is extremely dangerous. [T]he political, military, and economic rationales for the arms trade, upon scrutiny, seem inadequate in relation to the scale, pace, and consequences of the flow of weaponry. [T]he impression gained is one of more or less uncontrolled escalation which is exacerbating regional tensions and fueling local rivalries and instabilities, thereby increasing the risk of open conflict.

Thomas Ohlson, *Bulletin of Peace Proposals*[2]

Introduction

According to the Stockholm International Peace Research Institute (SIPRI), arms sales in the international system are again increasing after a hiatus in the early 1990s.[3] Like-minded analysts contend that "the arms trade is extremely dangerous" in that it aggravates regional tensions, promotes instability, and "increases the risk of open conflict."[4] Similarly, there are those who assert that arms sales "prolonged and escalated wars, resulting in more suffering and destruction."[5] However, others claim that weapons sales equalize military asymmetries between " 'have' and 'have not' countries" and create a security environment where "deterrence will be enhanced."[6] It follows that deterrence provides for less death and destruction. Contradictory claims such as these provide fruitful ground for scientific research. If increasing numbers of arms sales do indeed influence the outbreak and outcomes of wars, then there should be empirical evidence of this. In this chapter, we undertake such an empirical examination of the relationship between global arms transfers and war outbreak and war outcomes in the international system.[7]

While there are no truly systemic theories concerning arms transfers and war, this fact should not deter us from undertaking an examination at the global level.[8] In order to do so, we must assess the global implications of subsystemic theories. There are at least three distinct possibilities concerning the relationship between arms transfers and the historical outbreak of wars when aggregated at the systemic level. First, arms transfers may be early warning indicators of impending wars, and thus serve as preludes to war. To this logic, states may anticipate wars (for whatever reason) and begin arming prior to their outbreak.[9] Alternatively, there is the possibility that arms transfers relate to lesser amounts of war initiation via the creation of deterrence between rivals.[10] Inherent in this claim is the *para bellum* doctrine "If you want peace, prepare for war," which provides an alternative relationship to the one above. Finally, it is obvious that arms transfers often postdate termination of wars. In this case, the increasing levels of arms transfers in the international system are deemed a response to resupply needs of states involved in recently concluded wars, and not a causal predictor of war.

Besides questions surrounding arms transfers and war outbreaks, there are also largely uninvestigated implications of arms transfers on war outcomes, as noted by Pearson, Brzoska, and Crantz.[11] According to the SIPRI, "perhaps the most important questions about arms supplies" are their effect "on the course of wars and their general severity."[12] There are several possibilities of relationships between arms transfers and war outcomes. First, arms transfers may make wars bloodier by increasing the lethality of the rival forces. In this line of reasoning, increased militarization in the international system is related to increased severity of war. Conversely, others argue that arms transfers make wars less bloody by allowing the recipient side to surge to victory, thus ending the bloodshed. Arms sales may correlate with longer wars or shorter wars along much the same logic as above. Finally, they may increase the magnitude of wars by drawing suppliers into their clients' conflicts. Thus, at the global level, analysts have postulated that arms transfers relate in conflicting manners to such historical phenomena as wars under way, number of states involved in wars, and increased bloodiness of war. Further, if the *para bellum* doctrine is correct, then deterrence should affect more states. If so, peace should prevail, resulting in less war, fewer participants, and less severity and magnitude of war ongoing in the international system.

In this chapter, we examine hypotheses derived from the assertions above. In order to do so, we must first discuss the theoretical basis of each of the claims concerning arms transfers and war outbreaks and outcomes. Second, a research design is offered that explicates the hypotheses, methods, and variables with which we explore the evidence. Third, we analyze the data on arms transfers and wars on the systemic level for the 1950–1992 period. Finally, conclusions are drawn.

Arms Transfers and War Outbreaks: Theoretical Considerations

As noted by Baugh and Squires, the most often heard argument is that "arms transfers increase military capability, which in turn is interpreted by opponents as hostile intent," leading to increased likelihood of war.[13] This assertion, and a related consideration concerning forms of arms racing other than weaponization (e.g., dollar values of military and other aid or military spending), has been investigated by a number of scholars.[14]

Most of the above researchers focus on "military buildups" or "arms races," measured as military spending over time, in various incarnations of Richardsonian models, in order to make such assessments.[15]

According to this logic, when one side increases its military capability, either because it is dissatisfied with its present position in the international or regional political environment or because it recognizes that conflict with another state is likely in the future, opponents interpret its moves as hostile in design. This, in turn, leads the opponent to acquire weapons—thus, the "conflict spiral."[16]

A second possibility is that arms transfers, in a way similar to that attributed to military buildups by Wallace, make wars more likely.[17] The logic here is that rather than tension and propensity to conflict being inherent in the system (due to status inconsistency or some other attribute), the acquisition of weapons themselves produces the propensity to war. According to SIPRI, this happens because the acquisition of weapons provides leaders with additional foreign policy options, increases the visibility of military solutions to foreign policy problems, and avails politicians the opportunity to gain a return on their investment.[18] As stated in SIPRI:

> If a country has devoted a considerable quantity of resources to the military, and if those in power have devoted a good deal of time to military questions, then they will tend to consider [engaging in] military disputes, looking for some return on their investment of resources and time. When a border conflict arises ... military measures are more likely to be one of the possible courses of action.[19]

However, the fact is that it is difficult to determine whether arms transfers produce tensions or are their result. Singer calls this the "armament-tension dilemma."[20] Even though it is possible to alleviate difficulties in determining causal direction by certain statistical techniques, there still remains the need for caution in making the determination that arms transfers are the source of tension between states.

A second conceptualization of the effect of arms transfers on war outbreak reminds us of the *para bellum* dictate "If you want peace, prepare for war."[21] In this case, arms transfers may reduce the likelihood of war, and thus directly contrasts the above models.[22] According to this model, states acquire weapons in order to deter their potential opponents either through maintaining a military balance (according to those who believe that balance of power leads to peace) or by creating a relationship where one state is clearly stronger than its adversaries (supported by preponderance of power theories). In order to confirm this relationship, we should see that global levels of arms transfers associate with less war initiation.

The final conceptualization of the relationship between arms transfers and war outbreak is that the former typically take place *after* the latter.[23] Because many times international embargoes limit the weapons trade to belligerents involved in wars, the real opportunities for weapons sales may come even after termination of the war. Wars lead to arms transfers because the destructive processes of war make resupply necessary.

Arms Transfers and War Outcomes: Theoretical Considerations

The literature concerning the effect of arms transfers on war outcomes is relatively undeveloped compared to that involved in the study of war initiation. This reflects a trend in the international relations literature, which has been surprisingly unconcerned with questions of war outcomes.[24] In this chapter, we are concerned with outcomes of war in terms of several different factors. First, is the value of weapons sold in the international system positively correlated with the number of battle deaths suffered in wars? The logic here is that increasing numbers of weapons in the international system will not only increase the likelihood of war, but also increase its severity. States, for whatever reason, have armed and fought wars. Accordingly, the more weapons transferred in the system, the more blood will be spilled.[25]

A second proposition concerning the effect of arms transfers on the outcomes of wars is that they make wars last longer. According to this logic, states that are at war would quickly exhaust themselves in the absence of imported weapons, since for the 1950–1992 period, most states in the international system did not have the capacity to manufacture their own. Therefore, the ensuing relationship is that the more weapons sold in the international system, the fewer wars that will terminate in a given year. From this, we assume that the "backlog" of unresolved conflicts (conflicts that were artificially lengthened due to the effects of arms transfers) creates a conflict environment in the international systems where a greater number of nation-months of war exist. Subsequently, a greater percentage of possible war in the system (every state at war for the entire year) should be exhausted. By doing so, we can replace the probabilistic estimation of war termination with a more accessible formulation, that of global nation-months of ongoing war, percentage possible war levels exhausted or the raw number of wars under way.

Finally, the "resupply model" discussed above must be accounted for in terms of war outcomes. While there is no reason to believe that wars with more battle-related deaths, or more intense wars will result in higher levels of arms sales,[26] we expect that greater magnitudes of war will. Simply stated, the more states involved in war, the more likely that there will be a greater need for subsequent arms sales due to resupply needs.

Research Design: Statistically Surveying the Empirical Relationships between Arms Transfers and War Outbreak and Outcomes at the Systemic Level

Because, given the relations stated above, we would like to establish the causal relationship between arms transfers and war outbreak and outcomes in the international system, we must first examine the necessary conditions of a causal relationship.[27] The absence of empirical examination is startling for the first of these conditions: covariation. According to this condition, in order for a causal relationship to be assessed between two variables, systematic changes in one variable must accompany variation in the other. Baugh and Squires, in their empirical study of the hypothesis that arms transfers are related to war outbreak, using a sample of Third World interstate wars (1950–1965) and SIPRI data, find that "only about six percent of the total variance in wars" is accounted for in their model.[28] However, even this relationship is problematic, as will be discussed below.

Sylvan, in research on the relationship between military aid and conflictual behavior, finds that "sharp increases in military assistance tend to change decidedly the behavior of the recipient nation ... toward increased conflict and decreased cooperation."[29] Sherwin, however, sought to establish relations between arms transfers and cooperative and conflict activity across regions during the 1967–1976 period.[30] Using data from the U.S. Arms Control and Disarmament Agency (ACDA), he found that results were not consistent. Surprisingly, he also found that arms transfers more strongly related to *amount* of state activity in the international system (signing treaties, diplomatic exchanges, etc.) than either cooperative or conflictual actions. These tests explained about 28 percent of the variance (versus 14 percent of that concerning conflictual activity, which includes war).

Baugh and Squires also comment on the second condition required for the assessment of causation, that of temporal order. According to this precept, for one variable to cause covariation in another, it must precede it in time. The relationship between the variables of Baugh and Squires' contemporaneous causation model of arms transfers' increasing the likelihood of war outbreak is hedged with the following caveat, "The relationship ... is between wars and transfers that occur *in the same year*, and does not actually prove whether the transfers precede war, follow war, or occur during wars in the form of prompt resupply. Such detail is simply unavailable in our data" (emphasis in original).[31] Indeed, their accumulated transfers model, where they examine the strength of the effects of cumulative amounts of weapon imports over a five–year period prior to the outbreak of war, explains only about 2 percent of the variation of war incidence. Schrodt, on the other hand, finds that arming activities postdate conflictual behavior in his study of the Arabian Sea area from 1948 to 1978.[32]

The third condition for causal inference is that of nonspuriousness. In establishing the causal relationship between two variables, we must first determine that a third variable that is unobserved and therefore unaccounted for does not cause the variation between them. Baugh and Squires again attest to the difficulties in establishing this in the relationship between arms transfers and war. They admit that the weakness of their findings "should not come as a great surprise, given the many variables commonly believed to be associated with the outbreak of war."[33] In effect, their analysis reinforces the notion that the relationship between arms supply and war outbreak is not recursive, but is instead nonrecursive; i.e., arms transfers may be a causal factor in the outbreak of war, and the outbreak of war may cause arms transfers because of the need for resupply.

However, this logic apparently does not hold for questions of war outcomes—we have no reason to believe that bloodier wars (those with more battle deaths) cause arms transfers. The research above shows that the conditions for establishing causality between arms transfers and war outcomes are undeveloped. Because of this, we must establish the most basic aspects of this relationship in this chapter. In the past, some researchers eliminated data and analysis of war consequences. To reject the study of war consequences because of "space considerations and ... earlier analyses [which] proved theoretically uninteresting" is unfortunate in a discipline that values scientific and cumulative advancement of knowledge.[34] The stance taken throughout this research is that the consequences of war and the direct effect of weapons transfers thereon are important in both theoretical and practical terms.[35]

Research Hypotheses Derived for Arms Transfers and War Outbreak

Because of technical difficulties in satisfying the conditions necessary in establishing causality between related variables, we must rely on methods that are more appropriate for establishing correlations between them. Hypotheses derived from the discussion above concerning the effects of arms transfers on war outbreak reflect the exploratory nature of research on this topic at the international level.

While the effects of arms transfers on war have not been the subject of empirical examination at the systems level, clearly there are several ways in which these effects could manifest themselves. First, the absolute amount of the arms trade may correlate with an increase in war initiation. This reflects a simple, direct relationship between weapons and war. The second potential relationship encapsulates the idea that a single year of increased arming is unlikely to result in increased conflict initiation. However, if a researcher looks at the cumulative total of weapons acquired over time, higher averages over three- or five-year periods may associate with increases in war initiation, as indicated by Baugh and Squires.[36] Finally, it is possible that neither absolute nor cumulative increases affect the incidence of war, but sharp increases in the rate of arming do.[37] According to this logic, this is because there are qualitative as well as quantitative differences between "normal" military acquisitions and the less peaceful dynamics of "military buildups" or "arms races."

There are two additional hypotheses that reflect positive relationships between arms transfers and war outbreak examined in this chapter. First, as noted above, some scholars have found that arms transfers do not precede the outbreak of war, but rather war leads to arms transfers. Second, it may be argued that in addition to simple direct and positive relationships between arms transfers and war outbreaks, we should also see that there are "rank order" effects at work as well. In other words, years with the most arms transfers should correlate with those in which the most wars began.

- *Hypothesis 2-1a:* Increases in the amount of arms transfers (contemporaneous, one-, two-, and three-year lags) in the international system correlate positively with increased number of wars initiated in given years.

- *Hypothesis 2-2a:* Increases in the *cumulative amount* of arms transfers in the international system correlate positively with increased number of wars initiated in given years.

- *Hypothesis 2-3a:* Increases in the rate of change in arms transfers associate positively with positive increases in the number of wars begun in a given year.

- *Hypothesis 2-4a:* Increases in the number of wars begun in a given year associate positively with *subsequent* increases in the amount of arms transfers.

- *Hypothesis 2-5a:* Years with higher amounts of arms transfers will correlate with years with increased numbers of war outbreak.

Further, for the first three hypotheses above there are alternatives fit for the *para bellum* dictate (these are designated *Hypotheses 2-1b, 2-2b, 2-3b,* respectively). Here, arms transfers relate to lesser rates of war initiation because according to the *para bellum* argument, arming serves to preserve peace through deterrence. Hypotheses 2-4a and 2-5a do not require this alternative for theoretical reasons.[38] Finally, confirmation of the null hypotheses that declare that there is no relationship (i.e., either positive or negative) between arms transfers and war initiation is possible, as found by Milstein.[39]

Research Hypotheses Derived for Arms Transfers and War Outcomes

A second set of hypotheses derived from the theoretical assertions between arms sales and war outcomes follow. Again, it is important to investigate the three basic views on how arms sales should affect war;

i.e., yearly (and lagged one-, two- and three-years), cumulative, and rate-of-change effects, as well as the related *para bellum* hypotheses (2-6b, 2-7b, and 2-8b) and the null hypotheses.

- *Hypothesis 2-6a:* Increased amounts of weapons transfers (yearly, lags, cumulative, and rate of change) correlate positively with yearly war severity (battle deaths) when controlling for number of wars and number of participants.

- *Hypothesis 2-7a:* Increased amounts of weapons transfers (yearly, lags, leads, cumulative, and rate of change) correlate positively with greater numbers of conflicts ongoing in a given year.

- *Hypothesis 2-8a:* Increases in the levels of arms transfers (yearly, lags, leads, cumulative, and rate of change) in the international system correlate positively with the magnitude of war under way.

Methods: Pearson Product-Moment Correlation Coefficients and Spearman's Rank-Order Coefficients

Tests of the hypotheses offered above will be conducted by using the measure of association known as the Pearson Product-Moment Correlation Coefficient, or Pearson's *r*. The correlation coefficient is a statistical measure that describes the degree and direction of relation between variables, or their covariation.[40] If we find that relationships exist between arms transfers and war outbreaks and weapons sales and war outcomes, then we will at times express them in a more useful form by taking advantage of the relationship between the correlation coefficient and the coefficient of determination (R^2) of a bivariate regression equation. Rank-order coefficients are created by arranging data so that they are ordered from largest to smallest values.[41] By analyzing the data in this manner, we can determine how closely the years in which the most wars began correlate with years that had the highest levels of arms transfers.

Data

There are several useful variables to analyze the relationship between arms transfers and war outbreak and outcomes. The following section provides the reasoning behind the (several) operationalizations of arms transfers, war outbreaks, and war outcomes for analysis at the systemic level.

The main independent variable of this study is arms transfers. There are three institutions that publish arms transfer data of varying types: SIPRI, ACDA, and the United Nations.[42] Because the latter is a strictly voluntary compendium where member states submit—or all too often do not submit—arms sales transactions and is available for only the 1991–1996 time period, it is of little use here. ACDA data has the most inclusive coverage in terms of weapon types, including subsystems, etc., but does not give information on individual sales and is available for only the 1963–1996 period.[43] SIPRI data is an open-source compilation of transfers of major weapon systems and specifies numbers of units contracted and number of units delivered over the life of the contract. SIPRI data are available in both "raw form," which allows the researcher to reconstruct arms transfers by military item, and in constant dollars yearly from 1950 to the present (see figures 2.1, 2.2, and 2.3).[44]

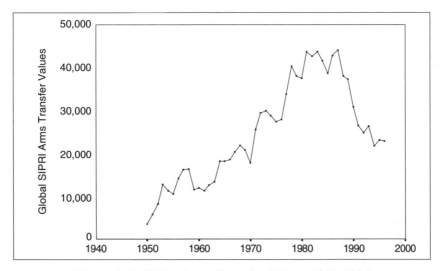

Figure 2.1: Global Arms Transfer Values, 1950–1996

The SIPRI data is not without limitations or controversy, as noted most clearly by Fei and Brzoska.[45] The two most important of these are the facts that its coverage is restricted to major weapons types (ships, artillery, missiles, armored vehicles, aircraft, and radar/guidance equipment) and that the information that is available is only that which can be gathered through open-source research (which was somewhat untrustworthy during the Cold War—especially for Soviet transfers to its allies and clients).[46] However, SIPRI's index of weapons value, which is unlike ACDA's normalized dollar values, represents a potentially superior means of measuring the effects of weapons sales. In essence, SIPRI maintains an index of weapons prices based on the constant dollar costs of producing the weapon (which includes research and development costs not covered by ACDA). Thus, when one country sells a weapon such as a main battle tank to another (any country, under any terms of contract, for any price), SIPRI categorizes it according to the index. This provides a useful means of evaluating the sale because it is not subject to the vagaries of the market.[47] Further, SIPRI data (1950–present) are available for a longer period of time than are ACDA data, thus allowing a more extensive exploration of the relationship between arms transfers and war outbreak and outcomes. For this chapter, SIPRI index values are used for the 1950 to 1992 period. These values, summed across states in the international system, provide yearly indicators of the arms trade. For the various tests, they are lagged, averaged, and recoded to rate-of-change values.

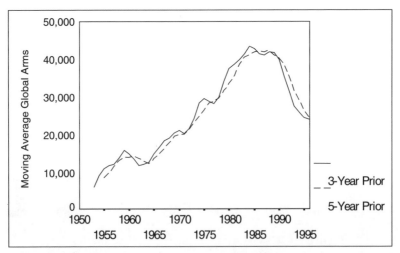

Figure 2.2: Moving Averages in Global Arms Transfer Values, 1950–1996

Data for the dependent variables, war outbreak, and outcomes come from the Correlates of War Project (COW) data sets. Included in the COW data is yearly information on the number of wars begun, participants, battle deaths, nation-months of war, and wars under way.[48] These data, and variations derived from them, encapsulate the dependent variables for this chapter (figures 2.5 through 2.10 contain time-series of these data).

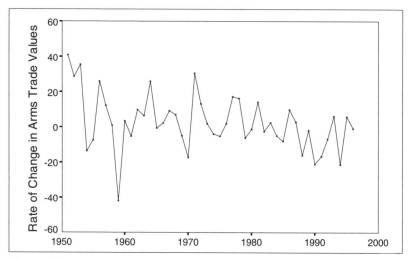

Figure 2.3: Rate of Change in Global Arms Transfer Values, 1950–1996

A few notes are in order given the peculiarities of the COW data.[49] COW codes wars based upon the criteria of 1,000 battle fatalities among all of the system members involved (to be included as a system member, a state must suffer either 100 battle deaths, or commit a minimum of 1,000 armed personnel to combat in the war). This admittedly is a rather conservative inclusion rule for coding wars, but such conservatism serves to make the empirical tests conducted here more rigorous.[50] The COW data for number of participants are available in two forms. First is information concerning the number of participants in wars that begin during the current year. Second is a figure for the participants in all ongoing wars during that year, i.e., wars begun and wars continuing. Likewise, COW provides battle deaths and nation-months of war indicators (the sum of all nations involved in war multiplied by the number of months each participant took part in the war during the year; i.e., 2 countries × 8 months = 16 nation-months of war). Finally, Small and Singer give various figures for amount of war under way for each year—in raw form (actual number of wars), normalized (number of participants divided by number of states in the international system), and a "percentage of nation-months exhausted." The latter is nothing more than the number of nation-months of actual war divided by the number of nation-months of possible war (i.e., every state in the system involved in 12 months of war for every year). Small and Singer assert that the percent exhausted measure is "perhaps the most sensitive and generally useful of the underway indicators."[51]

From the COW data described above, we construct the dependent variables needed to check correlations with arms transfers. Amount of war initiation per year is computed from the COW data on wars begun. This is the outbreak-dependent variable to be used in this chapter. Battle deaths, nation-months, and numbers of war per year provide us with the war outcome variables of severity, magnitude, amount of war, and duration (lessened probability of wars ending—built up over time). A final note: All of the war variables are combinations of the COW interstate and civil wars data for war itself, not the particulars of different types of wars (i.e., interstate, colonial, imperial, civil, etc.), interests us.[52]

Arms Transfers and War Outbreak and Outcomes: Some Empirical Evidence at the Systemic Level

The following analysis is broken into two parts. The first will deal with the relationship between arms transfers and war outbreak, while the second will cover arms sales and war outcomes.

War Outbreaks

Table 2.1 presents the Pearson correlations of Hypotheses 2-1a, 2-2a, and 2-3a. We must keep in mind that the alternative hypotheses derived from the *para bellum* thesis (Hypotheses 2-1b, 2-2b, and 2-3b) call for the opposite relationships. Thus, it is appropriate to report two-tailed tests, rather than one-tailed (which is more appropriate if we are able to predict the direction of the results). One final note: Because these tests cover the universe of states in the international system for the 1950–1992 period, inferential statistics are inappropriate. Significance levels serve only as a guide to substantive significance.

The results show that there is a positive, albeit modest, relationship between arms transfers and war outbreak on the systemic level. In terms of Hypothesis 2-1a, considering a direct (or lagged one to three years) relationship, we see that the Lag 1 variable has the highest correlation. This is important to note in terms of the problems discussed earlier concerning causal attribution—specifically temporal order. These findings *do not* indicate that we can safely attribute a causal link between arms transfers and war outbreak, even though the strongest relationship appears among those transfers that occur one year prior to higher numbers of war beginnings. Put simply, there is also a positive relationship between contemporaneous arms transfers and war outbreak.

Table 2.1: Correlates of Global Arms Transfers
and War Outbreak at the Systemic Level

Independent Variables (Cases) **Global Arms Transfers Values**	*Dependent Variable* **Total Wars Begun**
Hypothesis 2-1a (2-1b if negative)	
Contemporaneous (43)	0.2345
Lag 1 (42)	0.3021[*]
Lag 2 (41)	0.2513
Lag 3 (40)	0.2002
Hypothesis 2-2a (2-2b if negative)	
3-year Moving Average (40)	0.2210
5-year Moving Average (38)	0.1875
Hypothesis 2-3a (2-3b if negative)	
Rate of Change (42)	–0.2102
Rate of Change, Lag 1 (41)	–0.1798
Rate of Change, Lag 2 (40)	–0.0473
Rate of Change, Lag 3 (39)	–0.0364
Rate of Change, 3-year Moving Average (39)	–0.0708
Rate of Change, 5-year Moving Average (37)	–0.0482

* $p \leq 0.05$

Nonetheless, we can say that there are also positive relationships between three- and five-year cumulative buildups and war outbreaks (Hypothesis 2-2a). Again, these relationships are not strong, but they are in a direction such that we can refute the *para bellum* dictate in respect to building strong defense structures in order to prevent war. It appears that buildups associate with increasing numbers of war beginnings.

The results for Hypotheses 2-3a (2-3b) are surprising. The negative relationship between the rate of change of global arming and war outbreak indicates that arms races do not associate with war. However, the strength of the support provided here to the *para bellum* hypothesis is very weak to nonexistent—as exemplified by the very low (for the most part) correlations and their concomitant explained variance. The strongest variable associated with war outbreak, rate of change, only explains about 4 percent of the variance in the former. In substantive terms, these results indicate that there is a very weak association between global slowdowns in the weapons trade and outbreak of war.

The cumulative results of the test of global arms transfer values in relation to war outbreak support the claim that military buildups (in terms of arms transfers) lead to higher incidence of war. Unlike the previous findings of Baugh and Squires, however, the level of association between prior arms buildups (at least for one- and two-year lags) and war are higher than those of contemporaneous arms buildups and war.[53]

More evidence concerning the timing of arms transfers and war is available in Table 2.2. Here, we test the hypothesis that wars lead arms transfers; i.e., that wars begin prior to the influx of weapons.

**Table 2.2: Correlates of War Outbreak
and Global Arms Transfers at the Systemic Level**

Independent Variables (Cases) **Total Wars Begun**	*Dependent Variable* **Global Arms Transfers Values**
Hypothesis 2-4a	
0.2442	Lead 1 (43)
0.3069[*]	Lead 2 (43)
0.3031[*]	Lead 3 (43)

[*] $p \le 0.05$

When compared with the lag values from Table 2.1, we see that the lead values in Table 2.2 are slightly stronger. Thus, the more prominent effects between war outbreak and arms transfers is that the latter tend to follow the former. Again, this does not allow us to disregard the claims that arms transfers are an early warning indicator of war. However, it does allow us to temper this insight with the knowledge that a given increase in arms transfers is more likely the result of previous or ongoing wars, not the harbinger of additional conflict.

The final relationship to examine between arms transfers and war outbreak is that between the rank ordering. The findings for the hypothesis that years with higher amounts of arms transfers correlate with years of increased numbers of wars begun are summarized in Table 2.3. Lead values are included for comparative purposes.

The results in Table 2.3 indicate that there are weak relationships between the rank orderings of the arms transfers and war outbreak data. These findings are very similar to the Pearson's correlation tests, with war outbreaks leading arms transfers performing very slightly better than the lagged variables.

Table 2.3: Rank-Order Correlates of Global Arms Transfers and War Outbreak at the Systemic Level

Independent Variables (Cases) Global Arms Transfers Values	Dependent Variable Total Wars Begun
Hypothesis 2-5a	
Contemporaneous (43)	0.2090
Lag 1 (42)	0.2732
Lag 2 (41)	0.2237
Lag 3 (40)	0.1724
Lead 1 (43)	0.2352
Lead 2 (43)	0.3013[*]
Lead 3 (43)	0.2979

[*] $p \le 0.05$

War Outcomes

Table 2.4 provides the results from the examination of arms transfers on war outcomes. In this analysis, war severity (bloodiness) and lessened war termination as exemplified by greater numbers and magnitudes of conflicts under way interest us. The Pearson's correlation results of Hypotheses 2-6a, 2-7a, and 2-8a are presented; and again we should keep in mind that Hypotheses 2-6b, 2-7b, and 2-8b, supported by negative relationships, reflect the *para bellum* dictate that preparation for war will lead to peace (and therefore less magnitude of war ongoing, etc.).

In the test of Hypothesis 2-6a (2-6b) on the severity or bloodiness of wars, we find very weak negative correlations with the yearly and cumulative arms transfers variables when controlling for number of ongoing wars and participants. Likewise, when we look at the various rate of change indicators, we see such weak relationships that the explained variance of the best of these, rate of change on deaths per participant, is only a minuscule 3 percent. However, in tests not reported here, the rate of change variable had a slightly stronger correlation with battle deaths per nation-month ($r = 0.2508$). These associations sometimes indicate that weapons acquisition slowdowns over time associate with greater amounts of war deaths, but this relationship is very faint (only about 5 percent of the variance is explained by even the strongest relationship). Sharp increases in the rate of change of arms transfers generally have effects that are too weak and inconsistent to give much emphasis.

**Table 2.4: Correlates of Global Arms Transfer Values
and War Deaths**

Independent Variables (Cases)	*Dependent Variables*	
Global Arms Transfers Values	**Deaths/War**	**Deaths/Participant**
Hypothesis 2-6a (2-6b if negative)		
Contemporaneous (43)	–0.2221	–0.0595
Lag 1 (42)	–0.0835	–0.0661
Lag 2 (41)	–0.1440	–0.1141
Lag 3 (40)	–0.1493	–0.1163
3-year Moving Average (40)	–0.1578	–0.1330
5-year Moving Average (38)	–0.2265	–0.1976
Rate of Change (42)	0.0789	0.1809
Rate of Change, Lag 1 (41)	0.0246	–0.0877
Rate of Change, Lag 2 (40)	–0.0865	–0.1214
Rate of Change, Lag 3 (39)	0.1453	0.1288
Rate of Change, 3-year Moving Average (39)	0.0976	0.0089
Rate of Change, 5-year Moving Average (37)	0.0908	0.0057

When we move to Hypotheses 2-7a and 2-7b, concerning the amount and extent of wars under way, we find much stronger results (see Table 2.5). Yearly and cumulative arms buildups have strong, robust, positive associations with the number of wars under way and the number of participants in these wars, explaining about 58 to 91 percent of the variance. Interestingly, the effects—while strong in the direct relationships (contemporaneous and lags)—are even more prominent in the cumulative buildup of weapons over three- and five-year periods (see Figure 2.4). These findings lead us to conclude that there are discernible effects between arms transfers and wars. Arms transfers correlate with more wars under way, and subsequently they seem to affect the probability of war termination. At the same time, the rate of change variables reveal less powerful negative relationships with war outcomes (accounting for, at most, about 14 percent of the variance). As noted in the analysis of arms transfers and war outbreak above, this seems to indicate that arms races, or rapid and prolonged buildups over time, are not necessarily instrumental in influencing patterns related to war. Instead, there may be a lull before the storm.

**Table 2.5: Correlates of Global Arms Transfer Values
and War at the Systemic Level
(Amounts)**

Hypothesis 2-7a (2-7b if negative)	**Wars Under Way**	**Participants in Wars Under Way**
Contemporaneous (43)	0.8503***	0.7633***
Lag 1 (42)	0.9011***	0.7895***
Lag 2 (41)	0.9175***	0.8109***
Lag 3 (40)	0.9277***	0.8452***
Lead 1	0.8100***	0.7413***
Lead 2	0.7607***	0.7080***
Lead 3	0.6951***	0.6311***
3-year Moving Average (40)	0.9316***	0.8564***
5-year Moving Average (38)	0.9529***	0.9013***
Rate of Change (42)	–0.3145*	–0.1269
Rate of Change, Lag 1 (41)	–0.2699	–0.1338
Rate of Change, Lag 2 (40)	–0.2552	–0.1629
Rate of Change, Lag 3 (39)	–0.2263	–0.2520
Rate of Change, 3-year Moving Average (39)	–0.3336*	–0.3099*
Rate of Change, 5-year Moving Average (37)	–0.3702*	–0.2654

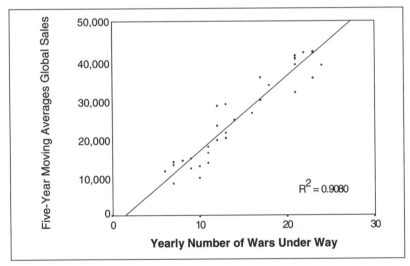

**Figure 2.4: Five-Year Prior Moving Averages of Arms Transfers
Predicting Amount of War Under Way, 1950–1992**

Table 2.5 also reflects the "resupply model," because it indicates that the number of wars under way does associate positively with subsequent levels of arms sales. The strength of this association seems to decline over time, as the Lead 1 variable has a stronger correlation with wars under way than does Lead 2, which is in turn more highly correlated than Lead 3. We see the same pattern, although a slightly weaker relationship, with the number of participants in wars under way in the same table.

Hypotheses 2-8a and 2-8b concern the relationship between arms transfers and the magnitude of wars under way. The results in Table 2.6 show that there are strong relationships between yearly and cumulative arms transfers and the number of nation-months of war under way and the percentage of possible nation-months of war exhausted in the system. Again, for each dependent variable, we see yearly arms transfers (and lagged transfers) having strong relationships, but that cumulative effects are greater. In this regard, whereas the weakest relationship in this group accounts for about 13 percent of the variance (between percent exhausted and the Lag 2 variable, $r = 0.3593$), close to 82 percent of nation-months under way is accounted for by five-year moving average ($r = 0.9074$)—the strongest. In the tests of these hypotheses, the rate of change variables had different relationships with the two dependent variables, but mostly very weak associations. The rate of change variables exhibited weak negative correlations with nation-months under way, but the explained variance is for the most part insignificant. However, the relationships between rate of change variables and the other dependent variable are mixed, with only one notable correlation (between rate of change and percent exhausted, which explains about nine percent of the variance).

When we turn to the lead variables that reflect the "resupply model," we again see strong positive correlations that weaken over time. Lead 1 is more highly correlated with the nation-months under way variable than Lead 2, which in turn is stronger than Lead 3. The same pattern holds for the relationship with percent exhausted, although these correlations are much weaker ($r = 0.4228$ for Lead 1 and percent exhausted compared to $r = 0.8090$ for Lead 1 and nation-months).

The overall results from the tests in Tables 2.4, 2.5, and 2.6 reveal that the strongest relationships between arms transfers and war outcomes are positive. These relationships indicated that yearly and cumulative amounts of arms sales strongly correlate with increases in war magnitudes (numbers of wars, participants, nation-months of war ongoing, and the exhausted amount of possible nation-months of war). Conversely, while the *para bellum* hypotheses received some support from the two-tailed tests, this support was inconsistent and relatively weak. Finally, we also found evidence to indicate that the relationship between war outcomes and arms transfers is recursive. Just as greater amounts of prior arms sales are strongly correlated with amount and magnitude of war, so are the latter related to subsequent arming patterns. While conclusive statements cannot be made based on the analysis performed here, the trends do indicate that while the effects of war on subsequent arming patterns decrease over time, the opposite is the case when we consider prior arms transfers. The strongest relationships are observed between cumulative prior sales over a five-year period and number of wars under way, number of participants in wars, nation-months of war under way, and the percent of possible nation-months in the system exhausted.

Table 2.6: Correlates of Global Arms Transfer Values
and War at the Systemic Level
(Magnitude)

Hypothesis 2-8a (2-8b if negative)	Nation-months Under way	Percent Exhausted
Contemporaneous (43)	0.8172^{***}	0.4114^{**}
Lag 1 (42)	0.7995^{***}	0.3593^{*}
Lag 2 (41)	0.8229^{***}	0.4692^{**}
Lag 3 (40)	0.8607^{***}	0.6447^{***}
Lead 1	0.8090^{***}	0.4228^{**}
Lead 2	0.7543^{***}	0.3817^{*}
Lead 3	0.6622^{***}	0.2928
3-year Moving Average (40)	0.8832^{***}	0.6798^{***}
5-year Moving Average (38)	0.9074^{***}	0.8073^{***}
Rate of Change (42)	–0.0351	0.3079^{*}
Rate of Change, Lag 1 (41)	–0.0903	0.1862
Rate of Change, Lag 2 (40)	–0.1466	0.0448
Rate of Change, Lag 3 (39)	–0.1795	–0.0824
Rate of Change, 3-year Moving Average (39)	–0.2347	–0.1074
Rate of Change, 5-year Moving Average (37)	–0.1806	–0.0403

* $p \leq 0.05$, ** $p \leq 0.01$, *** $p \leq 0.001$

Conclusion: On the Empirical Relationship
Between Arms Transfers and War Outbreak
and Outcomes on the Systemic Level

This chapter has presented empirical evaluations that have previously been absent in the literature concerning the relationship between arms transfers and war outbreak and outcomes. We examined the contending notions that "arms transfers are a factor in decisions to go to war ... and generally prolonged and escalated wars" and that arms transfers will allow "deterrence [to] be enhanced" and lead to peace.[54] From a review of the literature, hypotheses were constructed, variables for arms transfers and war outbreaks and outcomes were operationalized, and tests evaluated.

The empirical record shows that yearly and cumulative totals of arms sales in the international system are only weakly, but positively, related to the total amount of war outbreaks in the system (see Table 2.7). There is a slightly stronger relationship between war outbreak and subsequent arms sales than between prior arms sales and war outbreak. While these findings are still rather weak, the inclusion of civil wars and more data points seems to render stronger correlations than found in previous studies.

At the same time, weak and negative correlations exist between these same arms sales variables and the war outcomes variables for severity. Strong and consistent affinities exist between yearly and cumulative arms sales and war extent and magnitudes. Rate of change variables are, for the most part, correlated weakly and negatively with war outbreak and outcomes variables. In many cases, these relationships were so minuscule as to be difficult to interpret. Others reveal that these variables were inconsistent in their explanation of war variables, sometimes providing positive effects, while for the most part showing negative effects. Finally, the investigations of the "resupply model" reveal that arming effects from previous years have stronger correlations with greater amounts and magnitudes of war than do wars on subsequent arms sales. While these observations do not countervail the resupply model, they do indicate that the emphasis of this relationship is immediate; the strongest correlations are in the year immediately subsequent.

These results, in tests conducted at the systemic level, give support to those arguing that the sale of weapons has direct effects on wars. While we see only weak support for those who assert that weapons acquisitions correlate with increased incidence of wars, the evidence concerning war outcomes is strong. This is of deadly importance in an international environment marked by increased numbers of major weapons sales. If history and the tests run here are valid guides, this trend portends a moderate increase in numbers of war outbreaks, greater numbers of wars under way, an increase in numbers of participants in war, and a greater proportion of states in the international system concurrently at war (either externally or internally). These are truly sobering projections.

Table 2.7: Summary of Chapter 2 Hypotheses, Findings, and Recommendations

Hypothesis	Findings	Recommendation
Hypothesis 2-1a: Increases in the amount of arms transfers (contemporaneous, 1-, 2-, and 3-year lags) in the international system positively correlate with increased number of wars initiated in given years (*Hypothesis 2-1b if negative*).	Weak, positive relationship between global arms transfers and global war initiations.	Accept Hypothesis 2-1a.
Hypothesis 2-2a: Increases in the *cumulative amount* of arms transfers in the international system positively correlate with increased number of wars initiated in given years (*Hypothesis 2-2b if negative*).	Weak, positive relationship between global arms transfers and global war initiations.	Accept Hypothesis 2-2a.
Hypothesis 2-3a: Increases in the rate of change in arms transfers positively associate with positive increases in the number of wars begun in a given year (*Hypothesis 2-3b if negative*).	Weak, negative relationship between global arms transfers and global war initiations.	Accept Hypothesis 2-3b.
Hypothesis 2-4a: Increases in the number of wars begun in a given year are positively associated with *subsequent* increases in the amount of arms transfers.	Positive relationship between global war termination and global arms transfers.	Accept Hypothesis 2-4a.

Continued

**Table 2.7: Summary of Chapter 2 Hypotheses,
Findings, and Recommendations
(Continued)**

Hypothesis 2-5a: Years with higher amounts of arms transfers will correlate with years with increased numbers of war outbreaks.	Weak, positive rank order relationship between global arms transfers and global war initiations.	Accept Hypothesis 2-5a.
Hypothesis 2-6a: Increased amounts of weapons transfers (yearly, lags, cumulative, and rate of change) positively correlate with yearly war severity (battle deaths) when controlling for number of wars and number of participants (*Hypothesis 2-6b if negative*).	Mostly negative and weak relationships between global arms transfers and war severity.	Accept Hypothesis 2-6b.
Hypothesis 2-7a: Increased amounts of weapons transfers (yearly, lags, leads, cumulative, and rate of change) positively correlate with greater numbers of conflicts ongoing in a given year (*Hypothesis 2-7b if negative*).	Strong, positive relationships between yearly, leads, lags, and moving averages of global arms transfers and number of conflicts. Weaker negative relationships between rate of change in global arms transfers and number of conflicts.	Accept Hypothesis 2-7a; more investigation of 2-7b for rate of change in arms transfers needed.
Hypothesis 2-8a: Increases in the levels of arms transfers (yearly, lags, leads, cumulative, and rate of change) in the international system positively correlate with the magnitude of war under way (*Hypothesis 2-8b if negative*).	Strong, positive relationships between yearly, leads, lags, and moving averages of global arms transfers and magnitude of war under way. Weak and mixed relationships between rate of change in arms transfers and magnitude of war under way.	Accept Hypothesis 2-8a.

Notes

1. H.C. Engelbrecht and F.C. Hanighen, *Merchants of Death: A Study of the International Armament Industry* (New York: Dodd, Mead, 1934).

2. Thomas Ohlson, "Third World Arms Exporters—A New Facet of the Global Arms Race," *Bulletin of Peace Proposals* 13 (1982): 201–20.

3. Ian Anthony, Pieter Wezeman, and Siemon Wezeman, "The Trade in Major Conventional Weapons," in SIPRI, *SIPRI Yearbook 1996, Armaments, Disarmament and International Security* (Oxford: Oxford University Press, 1996), 463–536.

4. Ohlson, "Third World Arms Exporters," 1982.

5. Frederic Pearson, Michael Brzoska, and Christopher Crantz, "The Effect of Arms Transfers on Wars and Peace Negotiations," in SIPRI, *SIPRI Yearbook 1992, Armaments and Disarmament* (Oxford: Oxford University Press, 1992), 399.

6. James Foster, "New Conventional Weapons Technologies: Implications for the Third World," in *Arms Transfers to the Third World: The Military Buildup in Less Industrial Countries*, ed. Uri Ra'anan, Robert Pfaltzgraff, and Geoffrey Kemp (Boulder, CO: Westview Press, 1978), 77.

7. The classic works on arms trade at the systemic level are Robert Harkavy, *The Arms Trade and International Systems* (Cambridge, MA: Ballinger Publishing Company, 1975) and Edward Laurance, *The International Arms Trade* (New York: Lexington Books, 1992). The former provides the first theoretical, macrolevel analysis serving to relate mainstream topics in international relations theory such as alliance patterns, bloc polarization, ideological content of rivalries and distribution of power among leading nations, and patterns in the international arms trade. Harkavy thus uses international levels of weapons sales as a dependent variable and uses systemic level predictors in a model that explains transfer patterns across supplier "regimes." Laurance provides an updated variation on this investigation. The research conducted here does not draw directly from the theory and analysis of Harkavy and Laurance, yet retains the spirit of their seminal research via focus on explicitly stated dependent and independent variables, hypotheses, expectations, and models of arms trade phenomena. Where Harkavy and Laurance were interested in explaining patterns in the international arms trade, I am interested in a relationship left unresolved by their work: the impact of the arms trade on war. A more methodological debt is acknowledged to Edward Laurance and Ronald Sherwin, "Understanding Arms Transfers through Data Analysis," in *Arms Transfers to the Third World*, ed. Ra'anan, Pfaltzgraff, and Kemp, 1978, 87–105. Their work on "multiattribute utility scores" provides an important precursor to the analysis in Chapter 4.

8. See Ian Anthony, P. Wezeman, and S. Wezeman, "The Trade in Major Conventional Weapons," 1996, 463–536.

9. There are many arguments as to why states may anticipate war. "Realists" view war as inherent in the relations between states, as in Hans Morgenthau, *Politics Among Nations: The Struggle for Power and Peace* (New York: Knopf, 1948); Kenneth

Waltz, *Man, the State, and War: A Theoretical Analysis* (New York: Columbia University Press, 1959); and Kenneth Waltz, *Theory of International Politics* (Reading, MA: Addison-Wesley, 1979). Thus, statesmen must constantly prepare for it in order to provide security in a self-help system. Others, such as William Moul, "Predicting the Severity of Great Power War from Its Extent," *Journal of Conflict Resolution* 38 (1994): 160–69; and Michael Wallace, *War and Rank Among Nations* (Lexington, MA: D.C. Heath, 1973), contend that some states are dissatisfied with their status in the international system and therefore may import weapons in order to increase their power. Such imports cause "conflict spirals," which may lead to war, according to Randolph Siverson and Paul Diehl, "Arms Races, the Conflict Spiral, and the Onset of War," in *Handbook of War Studies*, ed. Manus Midlarsky (Boston: Unwin Hyman, 1989). Finally, some—such as SIPRI, *The Arms Trade With the Third World* (Stockholm: Almquist and Wiksell, 1971)—argue that weapons transfers increase the number of choices that decision makers have when considering foreign policy. In particular, they heighten the visibility and viability of militaristic options so that when disputes between states arise (which might otherwise be resolved peacefully), leaders are more confident of their own military prowess and thus more likely to go to war.

10. Foster, "New Conventional Weapons Technologies," 1978, argues that deterrence will lead to peace by virtue of the creation of military balances where none existed before through the managed transfer of "defensive" weapons. For more general treatments of deterrence theory and its role in preserving peace, see Ted Hopf, *Peripheral Visions: Deterrence Theory and American Foreign Policy in the Third World, 1965–1990* (Ann Arbor: University of Michigan Press, 1994); Paul Huth and Bruce Russett, "What Makes Deterrence Work?: Cases from 1900–1980," *World Politics* 36 (1984): 496–526; Paul Huth and Bruce Russett, "Deterrence Failure and Crisis Escalation," *International Studies Quarterly* 32 (1988): 29–45; Paul Huth and Bruce Russett, "Testing Deterrence Theory: Rigor Makes a Difference," *World Politics* 42 (1990): 466–501; Paul Huth and Bruce Russett, "General Deterrence between Enduring Rivals: Testing Three Competing Models," *American Political Science Review* 87 (1993): 61–73; Frank Zagare, *The Dynamics of Deterrence* (Chicago: University of Chicago Press, 1987); Frank Zagare, "Rationality and Deterrence," *World Politics* 62 (1990): 238–60.

11. Pearson, Brzoska, and Crantz, "The Effect of Arms Transfers on Wars and Peace Negotiations," 1992; see also Michael Brzoska and Frederic Pearson, *Arms and Warfare: Escalation, De-escalation, and Negotiations* (Columbia: University of South Carolina Press, 1994).

12. SIPRI, *The Arms Trade with the Third World*, 1971, 73.

13. William Baugh and Michael Squires, "Arms Transfers and the Onset of War Part I: Scalogram Analysis of Transfer Patterns," *International Interactions* 10 (1983): 39–63.

14. See SIPRI, *The Arms Trade With the Third World*, 1971; Arthur Burns, "A Graphical Approach to Some Problems of the Arms Race," *Journal of Conflict Resolution* 3 (1959): 326–42; J. David Singer, *Deterrence, Arms Control, and Disarmament* (Columbus: Ohio State University Press, 1962); J. David Singer, ed., *Correlates of War I: Research Origins and Rationale* (New York: Free Press,

1979); J. David Singer, ed., *Correlates of War II: Testing Some Realpolitik Models* (New York: Free Press, 1979); J. David Singer, ed., *Explaining War: Selected Papers from the Correlates of War Project* (Beverly Hills: Sage, 1979); John Stanley and Maurice Pearton, *The International Trade in Arms* (New York: Praeger, 1972); Michael Wallace, *War and Rank Among Nations* (Lexington, MA: D.C. Heath, 1973); Michael Wallace, "Arms Races and Escalation," *International Studies Quarterly* 26 (1982): 37–56; Michael Wallace, "Arms Races and Escalation: Some New Evidence," *Journal of Conflict Resolution* 23 (1982): 3–16; Paul Diehl, "Arms Races and Escalation: A Closer Look," *Journal of Peace Research* 20 (1983): 205–12; Paul Diehl, "Arms Races to War: Testing Some Empirical Linkages," *Sociological Quarterly* 96 (1985): 331–49; Paul Diehl and Jean Kingston, "Messenger or Message?: Military Buildups and the Initiation of Conflict," *Journal of Politics* 49 (1987): 801–13; Michael Altfield, "Arms Races?—and Escalation?: A Comment on Wallace," *International Studies Quarterly* 27 (1983): 225–31; Alex Mintz, "Arms Exports as an Action-Reaction Process," *The Jerusalem Journal of International Relations* 8 (1986): 102–13; Alex Mintz, "Arms Imports as an Action-Reaction Process: An Empirical Test of Six Pairs of Developing Nations," *International Interactions* 12 (1986): 229–43; Randolph Siverson and Paul Diehl, "Arms Races, the Conflict Spiral, and the Onset of War," in *Handbook of War Studies*, ed. Manus Midlarsky (Boston: Unwin Hyman, 1989); Eric Weede, "Arms Races and Escalation: Some Persisting Doubts," *Journal of Conflict Resolution* 24 (1980): 285–88.

15. The model posited by Lewis Richardson in 1939 is the best known and most influential model of an arms race—and has innumerable derivatives. It is based on the idea that states in an inherently conflictual environment will engage in a dynamic process of interaction in their acquisition of weapons (termed an arms race). This process can be summarized by the differential equation $M_a = dM_a/dt$, which is the rate of change in a state's (a) supply of weapons over time. In this model, M_a has three positive determinants: a "defense term," which is the stock of weapons of the opponent; a "fatigue term" that represents the economic and opportunity costs of arms racing; and the "grievance term" or all other factors (historical, cultural, institutional, etc.) that influence the arms race. See Lewis Richardson, "Generalized Foreign Politics," *British Journal of Psychology Monographs Supplement* 23 (1939); Lewis Richardson, "Could an Arms Race End without Fighting?" *Nature* (1951) 4274: 567–69; Lewis Richardson, *Arms and Insecurity* (Pittsburgh: Boxwood, 1960); Lewis Richardson, *Statistics of Deadly Quarrels* (Chicago: University of Chicago Press, 1960); and also Anatoli Rapoport, "Lewis F. Richardson's Mathematical Theory of War," *Journal of Conflict Resolution* 1 (1957): 249–304; Anatoli Rapoport, *Fights, Games and Debates* (Ann Arbor: University of Michigan Press, 1960); Michael Intriligator and Dagobert Brito, "Strategy, Arms Races, and Arms Control," in *Mathematical Systems in International Relations*, ed. J. Gillespie and Dina Zinnes (New York: Praeger, 1976); Michael Intriligator and Dagobert Brito, "Formal Models of Arms Races," *Journal of Peace Science* 2 (1976): 77–88; Michael Intriligator and Dagobert Brito, "Can Arms Races Lead to the Outbreak of War?" *Journal of Conflict Resolution* 28 (1984): 63–84; Michael Intriligator and Dagobert Brito, "Arms Races and Instability," *Journal of Strategic Studies* 9 (1986): 113–31; Michael Intriligator and

Dagobert Brito, "Richardsonian Arms Race Models," in *Handbook of War Studies*, ed. Midlarsky, 1989; Walter Isaard and Charles Anderton, "Arms Race Models: A Survey and Synthesis," *Conflict Management and Peace Science* 8 (1985): 27–98.

16. See Mintz, "Arms Exports as an Action-Reaction Process," 1986; and Mintz, "Arms Imports as an Action-Reaction Process: An Empirical Test of Six Pairs of Developing Nations," 1986.

17. Wallace, *War and Rank Among Nations*, 1973; Wallace, "Arms Races and Escalation," 1982; Wallace, "Arms Races and Escalation: Some New Evidence," 1982.

18. SIPRI, *The Arms Trade with the Third World*, 1971.

19. SIPRI, *The Arms Trade with the Third World*, 1971, 74.

20. J. David Singer, "Threat-Perception and the Armament-Tension Dilemma," *Journal of Conflict Resolution* 2 (1958): 90–105.

21. See Foster, "New Conventional Weapons Technologies," 1978; and Amelia Leiss with Geoffrey Kemp et al., *Arms and Local Conflict*, Vol. 2 of *Arms Transfers to Less Developed Countries* (Cambridge: MIT Center for International Studies, 1970). There is, of course, the chance that arms transfers have no effect on the initiation of war. This null hypothesis is discussed below.

22. See also Davis Bobrow, P. Terrence Hopmann, Roger Benjamin, and Donald Sylvan, "The Impact of Foreign Assistance on National Development and International Conflict," *Journal of Peace Science* 1 (1973): 39–60; and Philip Schrodt, "Arms Transfers and International Behavior in the Arabian Sea Area," *International Interactions* 10 (1983): 101–27, who find that arms transfers are associated with cooperative behavior.

23. Ronald Sherwin, "Controlling Instability and Conflict through Arms Transfers: Testing a Policy Assumption," *International Interactions* 10 (1983): 65–99; and Schrodt, "Arms Transfers and International Behavior," 1983.

24. See Melvin Small and J. David Singer, *Resort to Arms: International and Civil Wars, 1816–1980* (Beverly Hills: Sage, 1982). For exceptions in the broader literature, none of which considers arms transfer effects specifically, see Allan Stam, *Win, Lose, or Draw: Domestic Politics and the Crucible of War* (Ann Arbor: University of Michigan Press, 1996); D. Scott Bennett and Allan Stam, "The Duration of Interstate Wars, 1816–1985," *American Political Science Review* 90 (1996): 239–57; Pearson, Brzoska, and Crantz, "The Effect of Arms Transfers on Wars and Peace Negotiations," 1992; Claude Cioffi-Revilla, "On the Likely Magnitude, Extent, and Duration of an Iraq-UN War," *Journal of Conflict Resolution* 35 (1991): 387–411; Zeev Maoz, "Power, Capabilities, and Paradoxical Conflict Outcomes," *World Politics* 41 (1989): 239–66; A.F.K. Organski and Jacek Kugler, *The War Ledger* (Chicago: University of Chicago Press, 1980); and William Thompson, "The Consequences of War," *International Interactions* 19 (1993): 125–47.

25. Battle deaths are admittedly an imperfect means of assessing the "bloodiness" of wars, given that quite often the blood spilled during a conflict involves many

civilians as well as soldiers. An alternative is to use total deaths during the war, which would necessarily force an even more troublesome inaccuracy into our analysis: the fact that for virtually all wars of which there is a historical record, more deaths occur due to disease than battle. As no data source apparently exists that separates disease- and battle-related deaths by civilian and military personnel for the length of this study (1950–1992), we must rely on the sources of data at hand.

26. Great destruction may be accomplished with very little expenditure of weaponry because accuracy of delivery is a more important determinant of damage than destructive power. Because of this, it is unclear whether greater damage equates to greater expenditure of munitions or destruction of weapons and therefore a heightened market for postwar weapons or munitions sales akin to the resupply model.

27. See George Bohrnstedt and David Knoke, *Statistics for Social Data Analysis*, 3d ed. (Itasca, IL: F.E. Peacock, 1994); and David Dressler, "Beyond Correlations: Toward a Causal Theory of War," *International Studies Quarterly* 35 (1991): 337–55, for a discussion of causal assumptions.

28. Baugh and Squires, "Arms Transfers and the Onset of War Part II," 1983, 138.

29. Donald Sylvan, "Consequences of Sharp Military Assistance Increases for International Conflict and Cooperation," *Journal of Conflict Resolution* 20 (1976): 609–36.

30. Sherwin, "Controlling Instability and Conflict through Arms Transfers," 1983.

31. Baugh and Squires, "Arms Transfers and the Onset of War Part II," 1983, 138.

32. Schrodt, "Arms Transfers and International Behavior in the Arabian Sea Area," 1983.

33. Baugh and Squires, "Arms Transfers and the Onset of War Part II," 1983, 138.

34. Small and Singer, *Resort to Arms*, 1982, 116.

35. See also Thompson, "The Consequences of War," 1993, for similar views.

36. Baugh and Squires, "Arms Transfers and the Onset of War Part II," 1983.

37. Hans Rattinger, "From War to War to War: Arms Races in the Middle East," *International Studies Quarterly* 20 (1976): 501–31.

38. It makes little sense to investigate whether an increased number of war initiations precedes lesser amounts of international arms transfers for the study here. This is perhaps a fruitful area of research for those interested in the effects of international arms embargoes on war belligerents. The idea that rank correlations should exist between arms transfers and war outbreak reflects a desire to increase the substantive evidence for the relationships between these phenomenon. It does not, therefore, appear to warrant an alternative.

39. Jeffrey Milstein, "American and Soviet Influence, Balance of Power, and Arab-Israeli Violence" in *Peace, War, and Numbers*, ed. Bruce Russett (Beverly Hills: Sage, 1972).

40. Kenneth Hammond and James Householder, *Introduction to the Statistical Method: Foundations and Use in the Behavioral Sciences*, 3d ed. (New York: Knopf, 1967); Bohrnstedt and Knoke, *Statistics for Social Data Analysis*, 1994.

41. Hammond and Householder, *Introduction to the Statistical Method*, 1967.

42. Further analysis of the strengths and weaknesses of SIPRI and ACDA data can be found in Michael Brzoska, "Arms Transfer Data Sources," *Journal of Conflict Resolution* 26 (1982): 77–108; Edward Fei, "Understanding Arms Transfers and Military Expenditures: Data Problems," in *Arms Transfers in the Modern World*, ed. Stephanie Neuman and Robert Harkavy (New York: Praeger, 1979), 37–48; Edward Kolodziej, "Measuring French Arms Transfers," *Journal of Conflict Resolution* 23 (1979): 195–227. The best work on the UN register of conventional weapons transfers is Edward Laurance, *Arms Watch: SIPRI Report on the First Year of the UN Register of Conventional Arms* (Oxford: Oxford University Press, 1993).

43. ACDA provides aggregate dollar amounts imported by a country (or exported from a state) per year, and are available in their yearly publication, *Worldwide Military Expenditures and Arms Transfers*. This source is available in paper and electronic formats upon request from the ACDA, and some versions are also available from the International Consortium of Political and Social Research (ICPSR) at the University of Michigan. It is gathered for ACDA by the U.S. Central Intelligence Agency and the U.S. Defense Intelligence Agency.

44. SIPRI data are available from the organization in various forms, depending on the research needs of the particular project. They are produced as part of the *SIPRI Yearbook* series, and are also available (in some formats) via the World Wide Web. Finally, additional data may be attained by visiting SIPRI. For a description of the SIPRI data-collection and index evaluation guidelines, see SIPRI, *SIPRI Yearbook 1996, Armaments, Disarmament and International Security* (Oxford: Oxford University Press, 1996).

45. Brzoska, "Arms Transfer Data Sources," 1982; Fei, "Understanding Arms Transfers and Military Expenditures," 1979.

46. Additional potential problems with SIPRI data are that the database has much better information on what has been ordered rather than delivered, and its focus on major weapons systems. In regard to the former, when strategists consider their military situation, they have to take into account not only their current status, but also what "windows of opportunity" exist in the future. See James Morrow, "A Twist of Truth: A Reexamination of the Effects of Arms Races on the Occurrence of War," *Journal of Conflict Resolution* 33 (1989): 500–29, for a formal presentation of this logic. In part, I argue that this calculation is based on actual and projected arming patterns of rivals (this line of thought will be developed further in Chapter 4). The latter problem—the reliance on major weapons-systems data—brings out the potential pitfall of making conclusions based on a weapons trade that does not have as much relevance to the large number of "civil wars" in the 1945–1992 data; i.e., that the SIPRI dataset does not include the small arms that are predominantly used in these conflicts. There are two responses to this. SIPRI's response is to indicate that it believes that trends in major weapons-systems trade are highly correlated with the trade in small arms and is thus a useful proxy. While this may not be wholly

convincing, there is little, if any, reliable empirical evidence to shed light on this assertion. Finally, because there is no large-scale database on the global small-arms trade, we must make do with the data at hand while (hopefully) working to make better data available.

47. A hypothetical example may serve to illustrate market vagaries. In the ACDA formulation, hypothetical sales of two M-60A1 heavy tanks to two different buyers (by the same or different sellers) may have different codings based on their normalized dollar prices (one state, perhaps, was able to negotiate substantial offsets or somehow otherwise obtain a better deal and thus a lower per-unit price). Such incongruity introduces distortion into the ACDA database for the purposes for which it would be used here, because the M-60A1s are the same weapon system, and the per-unit prices obtained reflected not differences in lethality, but instead differences in the weapons market. Nor do such distortions reflect the local military conditions which the tanks operate in, necessarily (i.e., there is no reason to think that the tanks should have been worth more, necessarily, in the Middle East compared to South Asia). The same sales in the SIPRI formulation would be coded equivalently; i.e., both M-60A1s would have the same index score, and would thus more accurately reflect their equivalent lethality (and hence, their ability to influence the outbreaks or outcomes of wars).

48. Small and Singer, *Resort to Arms*, 1982, 118–22.

49. See Small and Singer, *Resort to Arms*, 1982; Singer, *Correlates of War I,* 1979; Singer, *Correlates of War II,* 1979; Singer, *Explaining War,* 1979; J. David Singer and Melvin Small, eds., *The Wages of War: 1816-1965, A Statistical Handbook* (New York: John Wiley and Sons, 1972); and J. David Singer and Paul Diehl, eds., *Measuring the Correlates of War* (Ann Arbor: University of Michigan Press, 1990).

50. COW project has also produced data on militarized disputes—conflicts *between states* that involve military forces, but which do not meet the criteria of war. However, to use this measure would support the idiom that intra- and interstate wars are distinct political phenomena. See Charles Gochman and Zeev Maoz, "Militarized Interstate Disputes, 1816-1976: Procedures, Patterns, and Insights," *Journal of Conflict Resolution* 28 (1984): 585–615, for more information on these data.

51. Small and Singer, *Resort to Arms*, 1982, 123.

52. For examples of studies that use combined interstate and intrastate (civil) conflict in the same variable, see Starr, "Revolution and War," 1994; Starr and Most, "Diffusion, Reinforcement, Geopolitics, and the Spread of War," 1980; Richardson, *Arms and Insecurity*, 1960; and Richardson, *Statistics of Deadly Quarrels*, 1960.

53. Baugh and Squires, "Arms Transfers and the Onset of War Part II," 1983.

54. Pearson, Brzoska, and Crantz, "The Effect of Arms Transfers on Wars and Peace Negotiations," 1992, 399; Foster, "New Conventional Weapons Technologies," 1978, 77.

2

Appendix 2
War Statistics: Additional Data

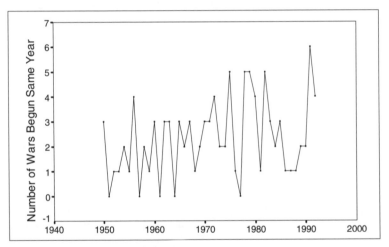

Figure 2.5: Number of Wars Begun, 1950–1992

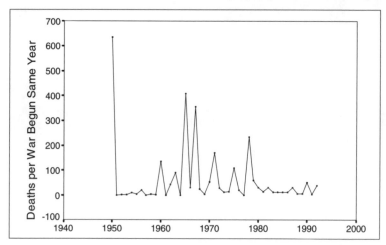

Figure 2.6: Number of Battle Deaths per War Begun, 1950–1992

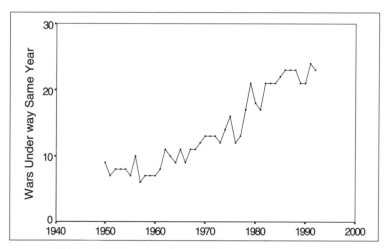

Figure 2.7: Number of Wars Under way, 1950–1992

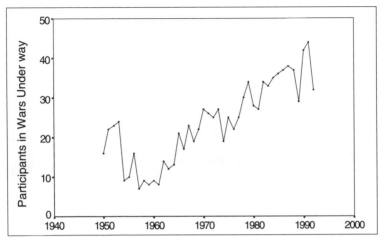

Figure 2.8: Number of Yearly Participants in War, 1950–1992

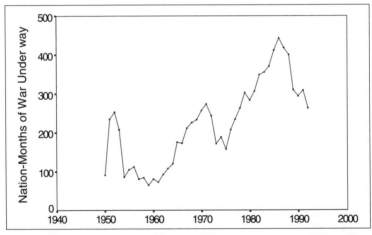

Figure 2.9: Total Nation-Months of War Under way, 1950–1992

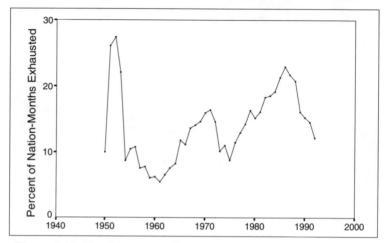

**Figure 2.10: Total Amount of Possible War in System Exhausted,
1950–1992**

3

The Effects of Arms Transfers on War Involvement and Outcomes, 1950–1992: Supplier and Recipient State Relationships

To give arms to all men who offer an honest price for them without respect of persons or principles: to aristocrats and republicans, to Nihilist and Czar, to Capitalist and Socialist, to Protestant and Catholic, to burglar and policeman, to black man, white man and yellow man, to all sorts and conditions, all nationalities, all faiths, all follies, all causes, and all crimes.

Creed of the arms maker, *Major Barbara*[1]

The Department of Defense has notified Congress that the Government of Egypt has requested the purchase of 32 AGM-84G (Block 1G) HARPOON missiles with containers, personnel training and training equipment, spare and repair parts, support equipment, publications, U.S. Government and contractor technical assistance and other related elements of logistics support. The estimated cost is $51 million .... This sale will contribute to the foreign policy and national security of the United States by helping to improve the security of a friendly country which has been and continues to be an important force for political stability and economic progress in the Middle East .... The sale of this equipment and support will not affect the basic military balance in the region.

U.S. Government Public Notification of Request for Weapons Sale[2]

Introduction

The egalitarian nature of the arms makers' creed seems perversely commendable, and the U.S. Department of Defense's regular assurance to Congress, and therefore the American people, that arms sales meet foreign policy needs and are not destabilizing provide a sense of studied precision. However, Ohlson and others make the more prominent assertion that increased globalization of the arms industry is threatening to peace and stability in the international system.[3] While the analysis from the preceding chapter indicates that weapons sales are directly tied to increased likelihood of wars begun, wars under way, and magnitude of wars in the international *system*, what is yet to be established is what effects arms sales have on individual *states*. It is argued that there are both supplier and recipient effects. For the suppliers, Pearson posits that arms transfers infer a strong commitment from supplier to recipient.[4] Subsequently, suppliers are more likely to intervene militarily in the affairs of recipients and thus more likely to become involved in war. For recipients, the assessment is that arms imports cause instability, arms races, military imbalances, and war.[5] Concurrently, there are alternative viewpoints. Arms sales and their more diffuse incarnation in the form of technology transfer may be seen as a benign form of business transaction, conducive to peaceful relations based on mutual deterrence between states, or at least neutral factors in war decision making.[6] To deprive companies, defense establishments, and countries of potential profits, employment, and beneficial production efficiencies represents an unnecessary encroachment on rights and privileges (however trade may be viewed in a given society) by the state. As espoused by the creed of the arms maker in Shaw's *Major Barbara*, arms merchants play a vital role in international relations by servicing a global, multibillion-dollar market. By doing so, they provide "economies of scale ... reduce research and development costs, reduce fixed costs, absorb nonrecurring costs, and create longer production runs or avoid gaps in production lines" that benefit society as a whole.[7]

While some excellent research has been conducted on the issues of supplier and recipient effects of arms transfers, most suffer from limitations incurred from constrained temporal periods, data inadequacies, and other methodological shortcomings.[8] It is the purpose of this chapter to update this research and provide additional empirical evidence concerning the effects of arms transfers on war involvement and outcomes so that we may more accurately assess whether global proliferation of conventional weapons portends a more dangerous world. Thus, the examination here, unlike the previous chapter where we examined the global arms trade, focuses on the effects of arms sales on the foreign policies of individual states. Specifically, do increased arms sales make conflict involvement more likely for either recipients or suppliers? Do arms sales to belligerents associate with bloodier and costlier conflicts? Are weapons acquisitions a means of increasing the likelihood of victory in war?

This chapter is organized as follows. First, a review of the literature concerning supplier and recipient effects of arms transfers is provided. Second, from this synthesis hypotheses are derived. This section explicates the models, methods, and data used. In the third section, the data are analyzed, and findings are presented. Finally, a summary and conclusion are offered concerning the theoretical and practical impact of the research presented.

Literature Review: The Effects of Arms Transfers on Suppliers and Recipients

Links between suppliers and recipients of arms transfers have been well described and analyzed.[9] In many of these studies, the researchers seek to trace linkages between arms transfers and supplier influence over recipient foreign policies. The logic is that states that face a military threat that they cannot meet through indigenous arming capability are dependent on supplier states. In turn, Quandt asserts that "it appears that decisions on military operations or policy concerning war and peace are the categories most likely to be influenced by an arms supplier."[10] Furthermore, dependency may create the greatest opportunity for outside influence when a state has only one major supplier. Yet, as noted by Catrina, "the search for empirical evidence of such an effect has so far proven difficult and less than wholly conclusive."[11]

If it is difficult to show that suppliers may influence the foreign policy behavior of recipients, then could it be that there is a "reverse dependence" where the former rely on the markets represented by the latter, and are in turn influenced by their policies? According to research conducted by Pearson, the answer is "yes."[12] Pearson examines whether arms transfers exemplify strong commitments to recipients and subsequently make the supplier more likely to intervene in the recipient's affairs. He finds that there is empirical evidence for such a pattern. However, supplier interventions are as likely to occur in states with multiple suppliers as those with only one, thus belying the expectations noted above.

The research findings of those who study recipient effects of arms transfers are more diverse. A rather large number of researchers have individually examined the question of whether arms transfers are correlated with increased conflictual behavior among recipients.[13] As noted by Kinsella, the findings of these studies are neither cumulative nor consistent.[14] There are those such as Bobrow and associates and Schrodt who find that arms transfers increase cooperative behavior, while others find the opposite.[15] There are also those who claim that there is no relationship, as well as some that conclude that there is no causality between transfers and conflict.[16] Another fertile area of research has been to examine the precise supplier-recipient relationship, i.e., the Cold War dynamics of U.S. and Soviet arms sales. These studies indicate that, at least in the Arab-Israeli dispute, American transfers probably did not exacerbate the conflict, but Soviet sales did.[17]

A second major question is whether arms transfers lead to arms races. Early work by Wright indicated that they do not.[18] However, empirical examination of this question reveals mixed findings. Rattinger, using a formulation of the classic Richardsonian arms race model,[19] found that arms transfers led to races prior to the 1967 Arab-Israeli War, but not prior to the 1973 war.[20] Mintz found that Soviet transfers reacted to Western sales, but saw no evidence for similar U.S. reactions.[21] In a different study, he found that in five of six cases, regional security dynamics followed the Richardsonian action-reaction process.[22]

The third question concerning recipient international behavior and its relationship with arms transfers revolves around the assessment of military capabilities. In groundbreaking studies, Sherwin and Laurance, Baugh and Squires, and Rattinger examine the effect of arms transfers on military capability.[23] Since this relationship does not directly impact the questions explored in this chapter, suffice it to say that these efforts represent important theoretical and methodological works for the conflict and arms trade literatures. Questions directly related to this research will be probed in the following chapter, which concerns the effect of arms transfers on war involvement and outcomes among competing contiguous dyads.

Finally, there are several important studies examining the effect of arms transfers on domestic conflicts. One of the main topics of research in this area explores the relationship between military expenditure and its implications for the socioeconomic structure and violence within recipient states. Building on Huntington's seminal work, Kemp and Miller argue that large amounts of imported weapons in underdeveloped countries potentially contribute to political instability.[24] There is reason to suppose that resources devoted to arms acquisition are necessarily diverted from other areas of the public sector. Such diversion is risky in many underdeveloped states that have great difficulty in promoting domestic tranquillity precisely because they are unable to deliver social goods.[25]

Of course, there is the possibility that arms imports are pursued precisely because of the previous instability in society. Several scholars assert that the majority of weapons imported into Third World states are used for repressing domestic opposition groups.[26] However, the relationship between arms imports and civil war remains somewhat ambiguous. Are elites able to suppress political dissent and even deter civil war by arming, or do weapons imports serve to increase the likelihood of rebellion? Empirical research reflects this uncertainty.[27]

Research Design: Conceptualization, Propositions, Empirical Models, and Data

The plethora of research reviewed above covers a large range of questions and propositions. For the purposes of this chapter, however, we are interested in only a few of these associations. First, we will confine our empirical examination to questions concerning arms transfers and war involvement and outcomes. In doing so, we choose a relatively difficult task, one made more challenging by the fact that war is—thankfully—a rare event in international relations. Its rarity should not be misconstrued. As noted by Ohlson and even more explicitly by SIPRI, "the most important question about arms supplies" is what effect they have "on the likelihood of wars breaking out, on the course of wars and on their general severity."[28] By focusing on these issues, we neglect considerations of the economic impact of arms transfers, supplier geopolitical dynamics and policies, and the effect of arms transfers on conflictual relations short of wars. This is not to imply that these issues are unimportant. They are. Our focus, however, results from the observation that empirical evidence concerning the effect of arms transfers on war involvement and outcomes remains either ambiguous or nonexistent.

A second and more particular arena of interest concerns supplier and recipient effects. As noted by one of the few studies that assesses the relationship of arms transfers and war onset directly, varying hypotheses formulate that arms transfers lead, lag, or are contemporaneous with wars.[29] From the supplier dimension, there are several possible relationships between arms transfers and wars. First, there could be very little impact on suppliers' foreign policies because in actuality their major role is to resupply the weapons that other states lose in civil or interstate wars. This "rebuilding model" is predicated on three likely scenarios: (1) to replenish the arms of a side that has lost a war; (2) to replace the losses of winners of wars; and (3) to restore a military balance in order to provide stability in a region where war has occurred.[30] Such scenarios are all based on the basic tenets of international relations theory, that all states must arm in order to provide for their own security and that states are sovereign within their borders. For the supplier, if another state wants to buy weapons, then it should sell them; each actor has the legitimate right to do so (if for defensive purposes). In a manner reminiscent of the arms maker's creed in Shaw's *Major Barbara,* supplier states service the market. The conceptualization of this relationship can be summarized as follows:

$$\text{Wars} \rightarrow \text{Need for Resupply} \rightarrow \text{Arms Transfers}[31] \qquad \text{(a)}$$

For the supplier, however, this conceptualization is potentially less benign than the representation above. According to Pearson, the supplier can get so heavily involved attempting to wield influence over the recipient state's policies that it in turn is drawn directly into conflicts (the classic example is, perhaps, the U.S. involvement in the Republic of Vietnam during the late 1950s, 1960s, and 1970s).[32] Thus, the effects of arms transfers for a supplier state on war involvement is likely to be positive; i.e., that increased amounts of arms sales portend an increased probability that it will become involved in war.[33] As noted by Gerner, this is indeed not a deterministic process, but nonetheless one that has been shown to exist empirically, albeit given a limited temporal span.[34] This conceptualization asserts, in the most basic formulation, that there is a direct relationship between arms transfers and supplier participation in war:

$$\text{Arms Transfers} \rightarrow \text{Increased Supplier Participation in Wars} \qquad \text{(b)}$$

Moving to recipient effects of arms transfers on wars, there are several additional considerations. While scholars and researchers tend to agree that there is some relationship between the receipt of weapons and war, there is much confusion over the nature of the relationship. The most common logic, according to Baugh and Squires, is as follows.[35] Arms transfers destabilize relations between states by altering the perception of military capabilities and evoking the conflict spiral that Jervis termed a security dilemma.[36] One state arms for purportedly defensive purposes, but other states have difficulty in assessing the first state's intentions; i.e., whether it might have aggressive aims. Thus, other states procure their own arms, which heightens the original weapons-buying state's feeling of insecurity, prompting it to buy even more weapons. The increased militarization in the relations among states increases the probability that war will occur between them because of the focus on military competitiveness, increased military readiness, and the prevalence toward military strategies and doctrines that favor the offensive over the defensive. All of these factors serve to make every state paranoid about its neighbor and the possibility of an attack. Baugh and Squires conceptualize this relationship as follows:

Arms Transfers→Military Capability→Security Dilemma→War[37] (c)

Alternatively, there are those who argue that arms transfers have an important effect on *preventing* war.[38] This argument encapsulates the classic Roman *para bellum* motto "If you want peace, prepare for war." In this case, arms transfers enhance military capabilities and convince each state that it cannot win in war. Because of the deterrent effect of increased military capabilities, peace ensues. This conceptualization also applies to theories that stress that a preponderance of power will lead to peace, because the weaker states will not want to fight against, and certainly be conquered by, the strong states:

Arms Transfers→Military Capability→Deterrence→Peace[39] (d)

In addition to the conceptualizations above, which emphasize the effect of arms transfers on war involvement, there are also alleged effects on war outcomes. According to Pearson, Brzoska, and Crantz, arms transfers generally make wars last longer and increase their severity and thus their costs.[40] This conceptualization is somewhat similar to the one above in that antebellum (or prewar) arms sales are assumed to even the military balance of contending states. This conceptualization contends that antebellum arms transfers lead to longer, bloodier, and costlier wars:

$$\text{Arms Transfers} \rightarrow \text{Military Balance} \rightarrow \text{Stalemate} \qquad \text{(e)}$$

However, there are alternatives to the above. Arms transfers could tip the antebellum military balance toward one side, thus allowing that side to emerge victorious. Another aspect of this theory states that arms transfers into an ongoing war may lead to war termination. By quick termination, wars will be less bloody and costly. What follows is a slight variation on Baugh and Squires' "Victory" conceptualization:

$$\text{Arms Transfers} \rightarrow \text{Military Imbalance} \rightarrow \text{War Outcomes}[41] \qquad \text{(f)}$$

The review of the literature above allows us to set up several propositions concerning the specific supplier and recipient effects of arms transfers on war outbreak and outcomes. These various conceptualizations imply sometimes conflicting predictions about the relationship of interest to us. For suppliers, we have the following propositions (the letter in parentheses indicates the corresponding model):

1. War predicts the "rebuilding" (supplier) dynamic of arms transfers after the termination of wars. (a)
2. Sales of weapons predict higher probability of war involvement among suppliers. (b)

For recipients, the subsequent propositions are:

3. Arms transfers predict increased recipient participation in war. (c)
4. Arms transfers predict decreased recipient participation in war. (d)

Finally, there are the following general propositions concerning the outcomes of wars:

5. Increased numbers of weapons sales prior to the onset of conflict predict longer (or shorter) wars. (e–f)
6. Weapons imports predict war termination. (f)
7. Increased numbers of weapons sales prior to the onset of conflict predict bloodier wars. (e)

8. Weapons imports prior to war outbreak predict less bloody wars. (f)
9. Antebellum arms imports predict war winners. (f)

In addition to the temporal sequences implied in the propositions above, we may also explore various lead and lag structures as well. Based on earlier works of Baugh and Squires, Kinsella, Kinsella and Tillema, and Sherwin, contemporaneous models are included with more temporally ordered ones (one-, two-, and three-year lags and leads; three- and five-year moving averages), although this will require some additional assumptions and manipulation of the arms transfer data as discussed below.[42]

The following models are used to test the propositions above. The models are numbered in association with the propositions above (e.g., 3–1 through 3–9) rather than with the conceptualizations examined earlier.

■ *Model 3-1:*

$$\text{MAVWARTERM} \Rightarrow \text{ARMSALES}_t ,$$

where MAVWARTERM = three- and five-year moving average of war terminations prior to time t; and ARMSALES_t = values of arms sales by four major and five minor post–World War II weapons suppliers at time t.[43] The major suppliers are: France, the United Kingdom, the United States, and the Union of Soviet Socialist Republics/Russia. As Russia inherited more than 90 percent of the Soviet weapons complex, the merger of Russian arms transfer values for the post-Soviet period is not problematic. The minor suppliers are: Israel, China, India, North Korea, and Czechoslovakia/Czech and Slovak Republics.

■ *Model 3-2:*

$$\text{ARMSALES}_{t, \text{ lags, MAVs}} \Rightarrow \text{SUPINWAR}_t ,$$

where $\text{ARMSALES}_{t, \text{ lags, MAVs}}$ = levels of arms sales by the above major and minor post-WWII suppliers (contemporaneous, one-, two- and three-year lags, and three- and five-year moving averages); and SUPINWAR_t = war involvement by supplier indicator (1 = involved, 0 = not involved).

■ *Model 3-3/4:*

$$\text{ARMSIM}_{t, \text{ lags, MAVs}} \Rightarrow \text{RECINWAR}_t \, ,$$

where $\text{ARMSIM}_{t, \text{ lags, MAVs}}$ = levels of arms imported by recipients (contemporaneous, one-, two- and three-year lags, and three- and five-year moving averages); and RECINWAR_t = war involvement by recipient indicator 1 = involved, 0 = not involved). Because of the oppositional nature of propositions 3 and 4, only one model is needed to test them both. If the relationship is positive, support for proposition 3 is in evidence. If the relationship is negative, proposition 4 is supported. The recipients are: Argentina, Egypt, Ethiopia, India, Indonesia, Iran, Iraq, Israel, Lebanon, Malaysia, Nicaragua, Pakistan, Somalia, Syria, Taiwan, and Uganda. For this model, we will control for the number of supplier states.

■ *Model 3-5:*

$$\text{ARMSIM}_{t, \text{ lags, MAVs}} \Rightarrow \text{WARLENGTH} \, ,$$

where $\text{ARMSIM}_{t, \text{ lags, MAVs}}$ is the same as above; and WARLENGTH is the duration of the war in nation-months.

■ *Model 3-6:*

$$\text{ARMSIM}_{t, \text{ lags, MAVs}} \Rightarrow \text{WARTERM} \, ,$$

where $\text{ARMSIM}_{t, \text{ lags, MAVs}}$ is the same as above; and WARTERM is an indicator for the year in which the war ends.

■ *Model 3-7/8:*

$$\text{ARMSIM}_{t, \text{ lags, MAVs}} \Rightarrow \text{WARDEATHS} \, ,$$

where $\text{ARMSIM}_{t, \text{ lags, MAVs}}$ is the same as above; and WARDEATHS is the number of battle-related deaths suffered by all belligerents during the war. Because of the oppositional nature of propositions 7 and 8, interpretation of the model will reveal which is supported by the data. If the relationship is positive, then the evidence backs proposition 7; otherwise, proposition 8.

■ *Model 3-9:*

$$\text{ARMSIM}_{t, \text{ lags, MAVs}} \Rightarrow \text{WARWINNER} \, ,$$

where $\text{ARMSIM}_{t, \text{ lags, MAVs}}$ is the same as above; and WARWINNER is an indicator variable for the winner of the war (1 = winner, 0 = not winner).

Statistical analysis of the above models is performed according to the dictates of the models. For those in which the response variable is binary, logistic regression is used. This technique is appropriate because (a) the dependent variable is dichotomous, while the independent variable (levels of arms transfers, either supplier or recipient) is interval-level; and (b) the relative incidence of war is low. Logistic regression uses maximum likelihood estimation (MLE) to estimate the unknown true population parameters.[44] In doing so, MLE performs successive estimations of the population parameters, finally accepting the set that yields the highest probability as the maximum likelihood estimates. MLE parameter estimates are unbiased for large samples, and are thus suitable for common statistical significance tests. There are four standard procedures for evaluating the overall fit of a logistic regression equation. First, the log likelihood ratio is an estimation of the improvement in fit brought to the model by including the predictor variables. Second, the goodness of fit statistic uses standardized residuals to compare observed probabilities to predicted values, and significance tests allow rejection of a null hypothesis that none of the logistic regression coefficients of the model differs significantly from zero. Third, the pseudo-R^2 is a descriptive measure that roughly estimates the amount of variation in the response variable accounted for by the predictors.[45] When used with caution, the pseudo-R^2 provides an assessment of the total performance of the explanatory model akin to the R^2 of ordinary least squares regression. Finally, one can classify each observation according to probabilities on the dependent variable by using logistic regression equations. Thus, given that observation's specific values for the predictor variables, comparisons can be made with the observed values for all N cases. Percentages of accurate classifications are used to assess the strength of the model in accounting for dichotomous outcomes (usually against a null model that predicts that all dichotomous values will be alike, either 1 or 0), and reductions in errors can be observed.

The remaining models have response variables measured at the interval level and can thus be analyzed by ordinary least squares (OLS) regression. As OLS is well understood within the discipline, we need not examine its technical details in this chapter.[46]

The data used in the empirical tests in this chapter come from two major databases. First, the Correlates of War (COW) project provides information about war outbreaks and outcomes. Included in this data set are facts concerning participants, dates of conflict initiation, and termination (from which war lengths are also derived), number of battle-related deaths, and winners. COW data concerning interstate, imperial, colonial, and civil wars are included with no distinctions between types in the tests conducted here, as we want to investigate the phenomenon of war—not its specific types. Wars, according to COW definition, only include military interactions in which there are at least 1,000 battle-related deaths per year for interstate, imperial, and colonial wars and 1,000 civilian and military deaths per year for civil wars.[47] While this is admittedly a stringent standard for war inclusion, it is one that is well accepted in the discipline; its conservatism serves to make analysis more rigorous.[48]

Second, SIPRI provides data concerning levels of arms transfers sorted according to suppliers and recipients. SIPRI data provides a credible source of data on transfers of six types of major weapon systems: armored fighting vehicles, aircraft, missiles, radar and guidance equipment, artillery systems over 100 mm, and ships. Whereas this data set does include information about the supply of (some) spare parts, research and development costs, depreciation of the product over time, and the new/refurbished/used status of the weapons, it does not include information on most "dual-use items," nor does it cover the small arms trade.[49] SIPRI's methodology attaches a monetary value to each piece of equipment in its data set based on a complex indexing system so that both quantitative and qualitative assessments of actual deliveries of weapons may be made.[50] SIPRI further assumes that major weapons constitute a fairly constant proportion of total military expenditures and thus provides a reasonable estimate of the flow of military equipment and assistance even without the inclusion of such items as small arms and training.

COW and SIPRI data were chosen for several reasons. First, they are both well accepted, respected, and widely used within the international relations and arms transfers communities. They are inclusive and based upon fairly well known and publicly available data and coding methods. Finally, SIPRI provides a longer temporal span (1950–1997) than its major competitor in the publication of arms transfers data, the U.S. Arms Control and Disarmament Agency [ACDA] (1961–1997), and is available in normalized figures (constant dollar equivalents). The temporal span of the data examined here is limited to the 1950 to 1992 period for arms transfers and covers 1945–1992 for wars. The 1992 cutoff is necessary due to the absence of a timely update of COW.

A final note on the data and models above is necessary. For the "contemporaneous" models, some data manipulation is necessary in order to assess the effects of arms transfers on war involvement and outcomes properly. As noted by Baugh and Squires, problems are inherent in contemporaneous models of this relationship because of the nature of yearly data.[51] Because one cannot determine precisely whether arms transfers in a given year preceded or postdated war outbreak or termination, there is difficulty in assessing causality. In this study, the following method is used for the empirical investigations. We will assess the level of arms imports or exports for a country in a year of war event (initiation or termination) by taking the proportion of the total value for the year (provided by SIPRI data) given the precise date of the event in relation to the year (from COW data). For example, if we are interested in a resupply model and a war terminated on 15 November 1966, a year in which $15 million of weapons were imported by country *a*, then the data would be recoded as $1.875 ($15 million multiplied by 0.125—the amount of the year still left after the end of the war). Similarly, if we were interested in the prewar buildup in a country that *began* a war on the same date, the data would be recoded as $13.125 million ($15 million multiplied by 0.875—the amount of the year already expired by 15 November). While this method of estimation is admittedly crude, it must suffice until further research is conducted concerning the precise arming patterns exhibited immediately before and after wars.[52]

Data Analysis: The Effect of Arms Transfers on War Outbreaks and Outcomes

The first model evaluated was Model 3-1. We searched for evidence to comply with the assertion that supplier states typically service a market for weapons created by the loss of weapons stocks due to war. This rebuilding model received very weak support in Baugh and Squires' research. There is also some reason to suspect that it holds due to the investigation performed in the previous chapter considering arms transfer effects on wars because of the finding that war outbreaks are associated with increased global arms transfers.[53] Before proceeding to the data analysis, some descriptive comments about the data are in order (see Figures 3.1, 3.2, and 3.3).

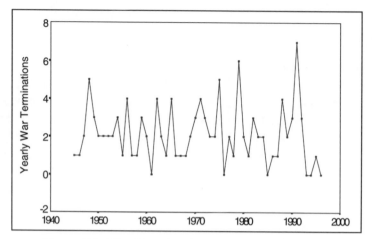

Figure 3.1: War Terminations per Year, 1945–1995

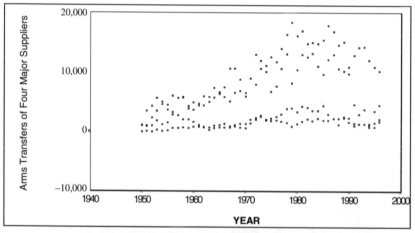

**Figure 3.2: Yearly Arms Transfer Values of the United States,
United Kingdom, France, and the Soviet Union/Russia, 1950–1996**

War terminations are nearly a yearly event in the international system (see Figure 3.1). Fortunately enough, 48 interstate wars and 63 civil wars have been brought to an end through a variety of peacemaking, peacekeeping, and conflict resolution methods—or due to termination via victory. Unfortunately, the encouraging frequency in conflict terminations does not necessarily portend a more peaceful world. However, frequent terminations may offer an attractive opportunity for arms suppliers looking to service the market for depleted military stocks without the complications of violating UN and other international embargoes against supplying weapons to belligerents.[54]

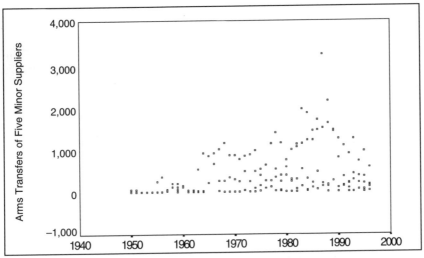

Figure 3.3: Yearly Arms Transfer Values for Czechoslovakia, North Korea, China, India, and Israel, 1950–1996

If this is the case, then one would expect to see a relatively distinct pattern of arms transfer values lagging behind war terminations. Figures 3.2 and 3.3 provide visual displays of the yearly arms transfer values for major suppliers (U.S., UK, France, and USSR/Russia) and minor suppliers (Czechoslovakia, North Korea, China, India, and Israel). These figures indicate that there is a persistent upward trend in arms transfer patterns.

Table 3.1 indicates, however, that there is barely any evidence that war *termination* precedes arms transfers as we would expect to see according to the resupply model. In fact, the findings presented below are weak enough to encourage us to reject proposition 1. The tests of Model 3-1 shows that there exists no generalizable relationship between prior three- and five-year averages of war termination and subsequent arms transfers. In 16 of the analyses (out of 22 iterations of the test), there is no statistical relationship discernible from zero (see Appendix 3, Tables A, B, and C, for the statistical results). Further, in five of the six tests that were significant, the direction of the relationship was the opposite of that contained in proposition 1.[55] In other words, *fewer* war terminations are predictors of increased arms transfers. This is particularly important preliminary evidence that indicates that the "resupply model" of arms transfers that take place after conflict termination is not well grounded empirically. Instead, supplier states seem to be servicing a market of a less benign sort—bringing to mind the creed of the arms maker in *Major Barbara*: to supply weapons "to all sorts and conditions, all nationalities, all faiths, all follies, all causes, and all crimes"—and perhaps doing so despite an international norm against supplying weapons to belligerents.

Table 3.1: Overview of Analysis of Proposition That War Termination Predicts Arms Transfers

Relationship with Three-Year Prior Moving Average	Sign	Relationship with Five-Year Prior Moving Average	Sign
Czechoslovakia	Negative	Czechoslovakia	Negative
India	Negative	United States	Positive
		USSR/Russia	Negative
		Minor Powers (Pooled)	Negative

Moving to the second proposition, that arms sales predict supplier war involvement and intervention, we find virtually no evidence to lend support to the findings of Pearson,[56] i.e., that previous arms transfers made suppliers more apt to intervene in the disputes of others. Model 3-2 attempts to predict the 101 cases of war involvement (of 349 cases) of the same four major arms suppliers and five minor suppliers used in tests above. Since war involvement represents only 28 percent of the cases, this is again a difficult task.[57] Table 3.2a provides the overall results of the pooled time-series logistic regression of this model.

In the MAV3 iteration, one correct "war involvement" prediction was obtained, thus distinguishing this iteration from the others, which were not different from the naive model. However, due to the low chi-square scores, we are encouraged to accept the null hypothesis that arms transfers are unrelated to supplier war involvement.

A logical extension of this model is to test whether it can detect the incidence of supplier intervention, i.e., the actual year in which the supplier intervened in the war. This is an even more difficult task than that above, because less than 10 percent of the cases (37 of 349) are interventions. Table 3.2b provides the results.

Table 3.2a: Arms Transfer Effects on Supplier War Involvement

Model	−2 log likelihood	Goodness of Fit	Chi-Square	Predicted (Improvement)	Pseudo -R²	Significance of B (sign)
Contemporaneous	416.190	349.388	3.735	71.06% (0%)	0	.05 (+)
LAG 1	400.240	340.387	4.470	71.76% (0%)	0	.03 (+)
LAG 3	371.234	323.564	5.132	73.07% (0%)	0	.02 (+)
LAG 5	343.673	305.544	5.263	74.10% (0%)	0	.02 (+)
MAV 3	340.111	298.520	6.593	73.49% (.33%)	.01	.01 (+)
MAV 5	315.861	278.725	6.302	73.38% (0%)	.01	.01 (+)

Table 3.2b: Arms Transfer Effects on Supplier War Intervention

Model	−2 log likelihood	Goodness of Fit	Chi-Square	Predicted (Improvement)	Pseudo -R²	Significance of B (sign)
Contemporaneous	235.927	349.034	.071	89.40% (0%)	0	.79 (−)
LAG 1	221.056	339.998	.000	90.00% (0%)	0	.99 (+)
LAG 3	208.349	322.782	.333	90.09% (0%)	0	.57 (−)
LAG 5	191.586	305.013	.040	90.49% (0%)	0	.84 (+)
MAV 3	181.124	297.974	.021	90.94% (0%)	0	.88 (+)
MAV 5	168.058	277.953	.060	91.01% (0%)	0	.80 (+)

Table 3.2b indicates that the various contemporaneous, lagged, and prior moving average iterations of Model 3-2 were able to provide no improvement over a naive model, which uniformly predicted "no war intervention." Again, the results warrant the acceptance of the null hypothesis that there is no relationship between arms transfers and supplier war intervention. The results concerning proposition 2 that arms sales make suppliers more likely to enter the conflicts of their client states receives little empirical verification by the tests conducted here. Thus, we are encouraged to reject it.

An examination of propositions 3 and 4 are provided to determine whether arms imports values predict war involvement by recipients. For recipients, tests were performed by using SIPRI yearly arms transfers values and numbers of suppliers (a control variable) and COW data for all wars from 1950 to 1992 that involved any of the following countries: Argentina, Egypt, Ethiopia, India, Indonesia, Iran, Iraq, Israel, Lebanon, Malaysia, Nicaragua, Pakistan, Somalia, Syria, Taiwan, and Uganda. Of these political entities, 13 were involved in civil wars, 13 participated in one or more interstate, colonial, and/or imperial wars, and 14 of the 16 were involved in some war (only Taiwan and Malaysia are coded by COW as no war involvement for 1950–1992). In 10 instances, a state was involved in more than one war simultaneously in a given year. While it was not intended when the sample was chosen, there is bias evident in this cohort. Even though in over 73 percent of all cases "no war involvement" is coded, the fact that nearly all of the states in the sample participated in some war during the period under examination provides a better than usual chance that increases in arms imports will predict wars. Model 3-3/4 provides the results given in Table 3.3a.

Table 3.3a: Arms Transfer Effects on Recipient War Involvement

Model	-2 log likelihood	Goodness of Fit	Chi-Square	Predicted (Improvement)	Pseudo-R^2	Significance of B (sign)
Contempor-aneous[58]	772.447	687.555	20.480	74.13% (.44%)	.00	.002 (+)
LAG 1	751.051	670.878	29.968	74.11% (.90%)	.02	.000 (+)
LAG 3	724.847	640.859	33.755	73.91% (1.88%)	.03	.000 (+)
LAG 5	706.974	608.691	28.228	72.20% (1.48%)	.03	.000 (+)
MAV 3	725.002	640.241	33.600	72.97% (.94%)	.03	.000 (+)
MAV 5	699.600	610.115	35.602	72.37% (1.65%)	.03	.000 (+)

These findings indicate that contemporaneous, lagged, and moving averages of arms import values are positive and statistically significant indicators of recipient war involvement. All of the time referent variations of this model refute the null hypothesis that none of the independent variables predicts war involvement, and they all provide small gains in predictive power over a naive model that provides a prediction estimate equivalent to the percentage of zeros (no war codings) in the sample. Thus, proposition 3 is upheld—that arms imports predict recipient war involvement, but with the caveat that the sample was slightly biased to favor that finding. Additionally, the evidence for proposition 3, especially given the sample bias, is weak. The "best" time-referent model, with three-year lagged import values, only predicts 13.41 percent of war involvement (24 of 179 cases) by the recipients (see Table 3.3b).

Table 3.3b: Classification Table
for Recipient War Involvement (Lag 3 model)

Observed	Predicted		Percent Correct
	No	Yes	
No	449	12	97.40%
Yes	155	24	13.41%
		Overall	73.91%

When it comes to the even more difficult task of predicting war outbreak (i.e., 1 is coded only in years of war initiation) involving arms importers, a similar model is unable to correctly predict the phenomenon and is, for the most part, indistinguishable from the naive model. The rarity of war outbreak cases, only 60 out of 688 cases, or 8.7 percent, provides a difficult challenge, but these models are logical extensions of proposition 3. Yet, in only two cases do any of the models differ from a naive model, and in these, the variable "Contemporaneous" predicts incorrectly while Lag 3 correctly predicts a single war outbreak (see Table 3.3c).

Table 3.3c: Classification Table
for Recipient War Entrance (Lag 3 model)

Observed	Predicted		Percent Correct
	No	Yes	
No	582	0	100.00%
Yes	57	1	1.72%
		Overall	91.09%

Even with the case selection unintentionally, but slightly, biased in favor of correct prediction of war involvement cases, analysts cannot correctly predict war involvement among arms transfer recipients by using these simple models. This, of course, has important ramifications for the normative aspects of conventional weapons control. However, before conclusively either accepting or rejecting propositions 3 and 4, more research should be conducted in order to determine whether the slight sample bias is unduly affecting the generalizability of these results, as unlikely as that may be.

The above tests concerned arms transfers and war involvement. The propositions explicated earlier indicated that there were both supplier and recipient effects in this relationship. The statistical analysis performed countervailed many of these expectations. Table 3.4 provides a summary of the propositions and the empirical results of their tests.

Table 3.4: Summary of Findings and Recommendations for Propositions 1 through 5 Concerning the Supplier and Recipient Effects of Arms Transfers on War Involvement

Proposition	Outcome	Recommendation
1. War predicts the "rebuilding" (supplier) dynamic of arms transfers.	Insignificance; wrong signs.	Reject Proposition 1.
2. Sales of weapons predict higher probability of war involvement among suppliers.	Insignificance; inability to improve on naive model.	Reject Proposition 2.
3/4. Arms transfers predict increased/ decreased recipient participation in war.	Some success in predicting *increased* recipient participation; failure to make substantively significant improvement over naive model in predicting war entrance.	Because of slightly biased sample (over-representation of war participation) and the weakness of the outcomes, more research is needed.

Whereas the research presented previously has shed new light on the phenomenon of arms transfer effects on war outbreaks, following is an examination of the effects of arms transfers on war outcomes. The first proposition (proposition 5 from page 58) tested concerning outcomes is that higher levels of antebellum weapons imports predict longer wars. Table D in Appendix 3 gives the statistical results of these models, which will be summarized here. For this test, a sample of 52 cases was used in which the previously named weapons importers participated in wars during the 1950–1992 era.[59] In short, none of the time-referent iterations of the model performed well enough to justify the rejection of the null hypothesis that there is no relationship between antebellum weapons imports and longer (or shorter) wars. A look at Figure 3.4 provides a visual aid in verifying that this conclusion is not confined to linear regression models over the 52-case sample.

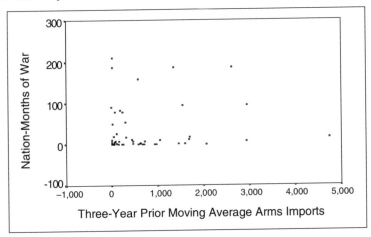

**Figure 3.4: War Length and Three-Year Moving
Averages of Weapons Imports**

Unfortunately for the scientific study of the effect of arms transfers and war outcomes, the relationship depicted in Figure 3.4 reveals that there may be no discernible pattern useful in predicting such effects.

The results from an examination of Model 3-5 immediately above informed us that there is no relationship between antebellum arms imports and war length. Model 3-6 explicates the proposition that antebellum weapons imports predict war termination. In this analysis, we can again use a pooled time series and logistic regression in order to conduct the test using the data on arms transfer recipients, 1950–1992. According to this hypothesis, the increased capabilities a belligerent accrues through importing weapons allow it to surge to victory. One would expect this to happen in a relatively short period of time, even provided that there may be a steep learning curve if the weapons imported are new, advanced, and/or both. Due to this expectation, the tests of this model investigated contemporaneous, one-, three-, and five-year lags, and three- and five-year prior (to war termination date) moving averages of weapons imports (see Table 3.5).

While predicting war termination in only 51 of 752 cases (about 6.8 percent) is admittedly an extraordinarily difficult task, it is clear from the analysis in Table 3.5 that we should reject the proposition that arms transfers predict war termination. In none of the temporal iterations could we be confident in rejecting the null hypothesis that there is no relationship between arms transfers and war termination based on the low chi-square scores. Further, because none of the temporal models was able to improve on the naive model that uniformly predicted no war termination, the models would be of little use even if they had met the rigors of statistical significance.

Table 3.5: Arms Transfer Effects on Recipient War Termination

Model	–2 log likelihood	Goodness of Fit	Chi-Square	Predicted (Improvement)	Pseudo -R²	Significance of B (sign)
Contemporaneous	368.061	744.851	4.872	93.22% (0%)	.01	.01 (+)
LAG 1	360.106	724.708	5.339	93.21% (0%)	.01	.01 (+)
LAG 3	358.114	698.485	2.724	92.90% (0%)	.00	.07 (+)
LAG 5	345.940	672.728	5.004	92.71% (0%)	.01	.01 (+)
MAV 3	356.178	696.115	4.659	92.90% (0%)	.00	.01 (+)
MAV 5	345.361	665.704	5.583	92.71% (0%)	.01	.01 (+)

Propositions 7 and 8 concern the effect on war bloodiness of antebellum transfers of weapons. Using the set of importers and completed wars from the test of proposition 5 above, tests were performed by ordinary least squares regression. The statistical results of these tests are available in Table E in Appendix 3 and are summarized here. In short, there appears to be no linear or linearizable pattern to the relationship between arms imports and subsequent battle deaths. The following figure provides visual evidence to confirm the statistical evidence of no relationship (see Figure 3.5 below).[60]

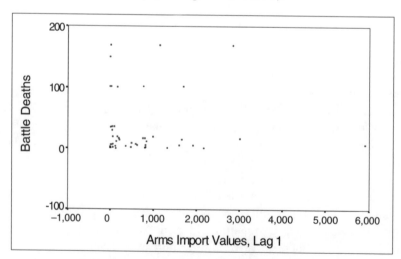

Figure 3.5: Relationship of Deaths to Arms Import Values, Completed Wars

Visual and statistical inspection of these data encourages the rejection of *both* propositions 7 and 8. It appears that there is no discernible pattern to indicate that arms imports make recipient war casualties either greater or lesser.

Investigation of proposition 9, that greater amounts of antebellum arms imports predict a higher probability of a belligerent emerging victorious in war, and Model 3-9 provided the results in Table 3.6. The data used, again from the sample of recipients that engaged in a war that began after 1950 and terminated prior to 1993, indicate that predicting this event, which occurred in 22 out of 52 cases, was not as difficult as in the tests above. Several of the iterations, Contemporaneous, LAG1, MAV3, and MAV5, were able to outperform the naive model. Although these improvements were only in the 2–7.5 percent range, these iterations also were able to reach a minimal level of statistical significance (at the .01 level) for chi-square.

Table 3.6: Arms Transfer Effects on Victory by Recipients in War

Model	−2 log likelihood	Goodness of Fit	Chi-Square	Predicted (Improvement)	Pseudo -R^2	Significance of B (sign)
Contemporaneous	68.024	51.723	2.828	59.62% (1.92%)	0.002	.14 (−)
LAG 1	58.508	45.015	10.596	64.71% (7.84%)	0.056	.01 (−)
LAG 3	66.788	50.139	2.316	58.82% (0%)	0	.20 (−)
LAG 5	66.146	50.085	1.156	60.00% (0%)	0	.32 (−)
MAV 3	60.481	46.135	8.624	60.78% (1.95%)	0.04	.02 (−)
MAV 5	63.576	49.805	3.726	62.00% (2.00%)	0.01	.10 (−)

Unfortunately, while the above LAG1 iteration of Model 3-9 reaches an acceptable level of statistical significance and also provides some overall improvement over a naive model, the relationship exhibited by the coefficient is opposite of that expected in accordance to proposition 9. Whereas proposition 9 predicts that greater amounts of weapons imports will provide victory in war, these tests show the opposite to be true. Instead, it appears that states that do not import vast amounts of weaponry are more likely to emerge victorious in war. Because of this, we are forced to reject proposition 9.

Table 3.7 summarizes the analysis provided here concerning arms transfers and their effects on war outcomes. Again, as with the previous analysis concerning war outbreak, the empirical evidence generated here indicates that the relationship between these two seemingly tightly associated phenomena are either unobservable, or very faintly observable.

**Table 3.7: Summary of Findings and Recommendations
for Propositions 5–9 Concerning the Supplier
and Recipient Effects of Arms Transfers on War Outcomes**

	Proposition	*Outcome*	*Recommendation*
5.	Increased numbers of antebellum weapons sales predict longer (or shorter) wars.	Insignificance; inconsistent signs.	Reject Proposition 5.
6.	Weapons imports predict war termination.	Insignificance; inability to improve on naive model.	Reject Proposition 6.
7/8.	Increased numbers of antebellum weapons sales predict bloodier/less bloody wars.	Insignificance; no discernible pattern to data.	Reject *both* Proposition 7 *and* Proposition 8.
9.	Antebellum arms imports predict war winners.	Mixed significance; wrong signs.	Reject Proposition 9.

Conclusion: The Supplier and Recipient Effects of Arms Transfers on War Involvement and Outcomes

The tests conducted in this chapter resulted from the allegations that arms transfers have distinct, measurable effects on supplier and recipient behavior concerning their decisions to enter or participate in wars, and on their conduct in those wars. The initial task was to review the theoretical and empirical literature so that propositions could be derived. From these propositions, models were created that allowed the empirical investigation of the purported relationships. Tables 3.4 and 3.7 summarize the findings for arms transfer effects on war outbreak and outcomes, respectively.

The findings indicate that most of the allegations, derived often from anecdotal evidence, concerning the effects of arms transfers on war are unfounded. Weapons sales by major and minor post–Second World War suppliers did not seem to follow a pattern consistent with the "rebuilding model." The supply of weaponry did not predict increased supplier involvement in conflict. Imports of weaponry did, however, prove to be related to recipient war involvement, but a combination of biased data and the weak predictive performance of the statistical model makes further investigation necessary to support the veracity and utility of these findings. Turning to the conduct of wars, the investigations of propositions 5 through 9 indicate that what little we thought we knew about the effect of weapons imports on war outcomes may be fallacious. Imports failed to predict longer wars, war termination, bloodier *or* less bloody wars, or war winners (although in the latter case some strong results containing the wrong signs were obtained, indicating the need for further investigation).

In short, the findings presented in this chapter are consistent with a meta null hypothesis of sorts: that arms transfers play little or no role in war involvement or outcomes for suppliers and perhaps even recipients as well. As a popular quip states, "Weapons don't make wars, men do." Indeed, the implications for efforts to control sales of military supplies, training, and technology are profound. The extant writings on weapons sales typified by Hartung or Wolpin are based on the normative estimate that war is of negative value and that weapons, because they are apparently necessary conditions for military operations, contribute to this negative value.[61] Investigations such as the present one are designed to determine empirically whether weapons sales do indeed contribute to the negative attributes of war. Such an understanding is necessary within peace research.

The findings obtained through these tests, however, inform us to different realities and in so doing open alternative explanations. Weapons sales are within the realm of various states' foreign policies. Perhaps there is little pattern regarding the use of arms transfers as a foreign policy tool across states. Further, even within the foreign policy "history" of a single state, there may be alternative explanations for weapons sales: to strengthen allies so that they may defend themselves; to extend deterrence to distant political friends; to bolster the flagging strength of a faltering ally; to prolong the destruction (and distraction) of a conflict that involves an enemy; to take advantage of market opportunities caused by destruction in wars. There are undoubtedly many others. If the investigations performed here withstand further research, the implication is that there is little or no pattern to use in the implicit (or explicit) categorization of the likely effects of any given weapons sale, even if the foreign policy goal of the supplier is known.

In the context of the world's largest supplier of military weapons, technology, and services, the U.S. government's notice to Congress regarding the benign nature of upcoming, policy-directed proliferation of high-tech weaponry around the world is misleading. To state that "[t]his sale will contribute to the foreign policy and national security of the United States by helping to improve the security of a friendly country which has been and continues to be an important force for political stability and economic progress" remains contentious.[62] There seem to be few, if any, patterns by which we could make such predictions. In the next chapter, we will investigate the subsequent assertion that "the sale of this equipment and support will not affect the basic military balance in the region."[63]

Notes

1. From George Bernard Shaw's play *Major Barbara*, produced in 1905.

2. "Memorandum for Correspondents," No. 156-M. 3 September 1997. Provided by DefenseLINK and available at: [http://www.dtic.mil/defenselink/].

3. Thomas Ohlson, "Third World Arms Exporters—A New Facet of the Global Arms Race," *Bulletin of Peace Proposals* 13 (1982): 201–20. See also Richard Bitzinger, "The Globalization of the Arms Industry: The Next Proliferation Challenge," *International Security* 19 (1994): 170–98; Michael Brzoska and Frederic Pearson, *Arms and Warfare: Escalation, De-escalation, and Negotiations* (Columbia: University of South Carolina Press, 1994); Christian Catrina, "Main Directions of Research in the Arms Trade," *Annals of the American Academy of Political and Social Science* 535 (1994): 190–205; Peter Mason, *Blood and Iron: Breath of Life or Weapon of Death?* (Victoria, Australia: Penguin Books, 1984); Frederic Pearson, Michael Brzoska, and Christopher Crantz, "The Effect of Arms Transfers on Wars and Peace Negotiations," in SIPRI, *SIPRI Yearbook 1992, Armaments and Disarmament* (Oxford: Oxford University Press, 1992); George Seldes, *Iron, Blood and Profits: An Exposure of the World-Wide Munitions Racket* (New York: Harper and Brothers, 1934); and SIPRI, *The Arms Trade With the Third World* (Stockholm: Almquist and Wiksell, 1971).

4. Frederic Pearson, "An Analysis of the Linkage between Arms Transfers and Subsequent Military Intervention," Center for International Studies, Occasional Paper 8102, 1981.

5. See SIPRI, *The Arms Trade with the Third World*, 1971; John Stanley and Maurice Pearton, *The International Trade in Arms* (New York: Praeger, 1972); and Donald Sylvan, "Consequences of Sharp Military Assistance Increases for International Conflict and Cooperation," *Journal of Conflict Resolution* 20 (1976): 609–36.

6. See, respectively, Anne Schwarz, "Arms Transfers and the Development of Second-Level Arms Industries," in *Marketing Security Assistance: New Perspectives on Arms Sales*, ed. David Louscher and Michael Salamone (Lexington, MA: Lexington Books, 1987), 101–30; and David Louscher and Michael Salamone, *Technology Transfer and U.S. Security Assistance: The Impact of Licensed Production* (Boulder, CO: Westview Press, 1987); David Louscher and Michael Salamone, "The Imperative for a New Look at Arms Sales," in *Marketing Security Assistance,*

ed. Louscher and Salamone, 1987, 13–40. James Foster, "New Conventional Weapons Technologies: Implications for the Third World," in *Arms Transfers to the Third World: The Military Buildup in Less Industrial Countries*, ed. Uri Ra'anan, Robert Pfaltzgraff, and Geoffrey Kemp (Boulder, CO: Westview Press, 1978), 65–84; and Quincy Wright, *A Study of War*, two vols. (Chicago: University of Chicago Press, 1942) agree that "weapons do not make war" but rather men do.

7. Lewis Snider, "Do Arms Exports Contribute to Savings in Defense Spending? A Cross-Sectional Pooled Time Series Analysis," in *Marketing Security Assistance*, ed. Louscher and Salamone, 1987, 41–64.

8. Much of the arms transfer literature concerns supplier "models" or rationales for selling weapons abroad and the domestic and international economic effects of arms sales. Neither issue will be covered in this chapter. For excellent reviews of this literature, as well as theoretical and empirical treatment, see Stephen Kaplan, "U.S. Arms Transfers to Latin America, 1945-1974," *International Studies Quarterly* 19 (1975): 399–431; Paul Hammond et al., *The Reluctant Supplier: US Decisionmaking for Arms Sales* (Cambridge, MA: Oelgeschlager, Gunn and Hain, 1983); Christian Catrina, *Arms Transfers and Dependence* (New York: United Nations for Disarmament Research, 1988); Emile Benoit, *Defense and Economic Growth in Developing Countries* (Lexington, MA: Lexington Books, 1973); Luigi Einaudi et al., *Arms Transfers to Latin America: Toward a Policy of Mutual Respect*, Report R-1173-DOS, Santa Barbara, CA: Rand, 1973; Peter Frederikson and Robert Looney, "Defense Expenditures and Economic Growth in Developing Countries: Some Further Empirical Evidence," *Journal of Economic Development* 7 (1982): 113–26; Ulrich Albrecht, Dieter Ernst, Peter Lock, and Herbert Wulf, "Armaments and Underdevelopment," *Bulletin of Peace Proposals* 5 (1974): 173–85; Stephanie Neuman, "Arms Transfers and Economic Development: Some Research and Policy Issues," in *Arms Transfers in the Modern World*, ed. Stephanie Neuman and Robert Harkavy (New York: Praeger, 1979); Stephanie Neuman, "International Stratification and Third World Military Industries," *International Organization* 38 (1984): 167–97; Jan Oberg, "Third World Armament: Domestic Arms Production in Israel, South Africa, Brazil, Argentina and India," *Instant Research on Peace and Violence* 5 (1975): 1050–75; Gregory Sanjian, *Arms Transfers to the Third World: Probability Models of Superpower Decisionmaking* (Boulder, CO: Lynne Rienner, 1987); Gregory Sanjian, "Arms Export Decision-Making: A Fuzzy Control Model," *International Interaction* 14 (1988): 243–65; Gregory Sanjian, "Fuzzy Set Theory and U.S. Arms Transfers: Modeling the Decision-Making Process," *American Journal of Political Science* 32 (1988): 1018–46; Gregory Sanjian, "Great Power Arms Transfers: Modeling the Decision-Making Processes of Hegemonic, Industrial, and Restrictive Exporters," *International Studies Quarterly* 35 (1991): 173–93; Herbert Wulf, "Dependent Militarism in the Periphery and Possible Alternative Concepts," in *Arms Transfers in the Modern World*, ed. Neuman and Harkavy, 1979.

9. See, for example, Catrina, *Arms Transfers and Dependence*, 1988; Christian Catrina, "Main Directions of Research in the Arms Trade," *Annals of the American Academy of Political and Social Science* 535 (1994): 190–205; William Hartung, *And Weapons for All: How America's Multibillion-Dollar Arms Trade Warps Our Foreign Policy and Subverts Democracy at Home* (New York: HarperCollins,

1994); Keith Krause, "Military Statecraft: Power and Influence in Soviet and American Arms Transfer Relationships," *International Studies Quarterly* 35 (1991): 313–36; Pearson, "An Analysis of the Linkage between Arms Transfers and Subsequent Military Intervention," 1981; Christopher Shoemaker and John Spanier, *Patron-Client State Relationships: Multilateral Crises in the Nuclear Age* (New York: Praeger, 1984); and Lewis Snider, "Arms Transfers and Recipient Cooperation with Supplier Policy Preferences," *International Interaction* 5 (1979): 241–66.

10. William Quandt, "Influence through Arms Supply: The US Experience in the Middle East." In *Arms Transfers to the Third World: The Military Buildup in Less Industrial Countries*, ed. Uri Ra'anan, Robert Pfaltzgraff, and Geoffrey Kemp (Boulder, CO: Westview Press, 1978), 129.

11. Catrina, "Main Directions of Research in the Arms Trade," 1994, 202. See also Uri Ra'anan, "Soviet Arms Transfers and the Problem of Political Leverage," in *Arms Transfers to the Third World,* ed. Ra'anan, Pfaltzgraff, and Kemp, 1978, 131–56.

12. Pearson, "An Analysis of the Linkage between Arms Transfers and Subsequent Military Intervention," 1981. See also Miles Wolpin, *America Insecure: Arms Transfers, Global Interventionism, and the Erosion of National Security* (London: McFarland and Company, 1991); Hartung, *And Weapons for All,* 1994; and Catrina, *Arms Transfers and Dependence,* 1988.

13. See Ronald Sherwin, "Controlling Instability and Conflict through Arms Transfers: Testing a Policy Assumption," *International Interactions* 10 (1983): 65–99; Philip Schrodt, "Arms Transfers and International Behavior in the Arabian Sea Area," *International Interactions* 10 (1983): 101–27; Donald Sylvan, "Consequences of Sharp Military Assistance Increases for International Conflict and Cooperation," *Journal of Conflict Resolution* 20 (1976): 609–36; Stephanie Neuman, *Military Assistance in Recent Wars* (New York: Praeger, 1986); David Kinsella, "Conflict in Context: Arms Transfers and Third World Rivalries during the Cold War," *American Journal of Political Science* 38 (1994): 557–81; David Kinsella and Herbert Tillema, "Arms and Aggression in the Middle East," *Journal of Conflict Resolution* 39 (1995): 306–29; William Baugh and Michael Squires, "Arms Transfers and the Onset of War Part II: Wars in Third World States, 1950–65," *International Interactions* 10 (1983): 129–41.

14. Kinsella, "Conflict in Context," 1994.

15. Davis Bobrow et al., "The Impact of Foreign Assistance on National Development and International Conflict," *Journal of Peace Science* 1 (1973): 39–60; Schrodt, "Arms Transfers and International Behavior in the Arabian Sea Area," 1983. See also Kinsella, "Conflict in Context," 1994, in the Iran case. Baugh and Squires, "Arms Transfers and the Onset of War Part II," 1983; Sherwin, "Controlling Instability and Conflict through Arms Transfers," 1983; Sylvan, "Consequences of Sharp Military Assistance Increases for International Conflict and Cooperation," 1976.

16. Jeffrey Milstein, "American and Soviet Influence, Balance of Power, and Arab-Israeli Violence," in *Peace, War, and Numbers*, ed. Bruce Russett (Beverly Hills: Sage Publications, 1972); Baugh and Squires, "Arms Transfers and the Onset of

War Part II," 1983; Sherwin, "Controlling Instability and Conflict through Arms Transfers," 1983.

17. Kinsella, "Conflict in Context," 1994; Kinsella and Tillema, "Arms and Aggression in the Middle East," 1995.

18. Quincy Wright, *A Study of War,* 1942.

19. For reviews and analysis of Richardsonian arms race models, see Anatoli Rapoport, "Lewis F. Richardson's Mathematical Theory of War," *Journal of Conflict Resolution* 1 (1957): 249–304; Michael Intriligator and Dagobert Brito, "Richardsonian Arms Race Models," in *Handbook of War Studies,* ed. Manus Midlarsky (Boston: Unwin Hyman, 1989); and Walter Isaard and Charles Anderton, "Arms Race Models: A Survey and Synthesis," *Conflict Management and Peace Science* 8 (1985): 27–98.

20. Hans Rattinger, "From War to War to War: Arms Races in the Middle East," *International Studies Quarterly* 20 (1976): 501–31.

21. Alex Mintz, "Arms Exports as an Action-Reaction Process," *The Jerusalem Journal of International Relations* 8 (1986): 102–13.

22. Alex Mintz, "Arms Imports as an Action-Reaction Process: An Empirical Test of Six Pairs of Developing Nations," *International Interactions* 12 (1986): 229–43.

23. Ronald Sherwin and Edward Laurance, "Arms Transfers and Military Capability: Measuring and Evaluating Conventional Arms Transfers," *International Studies Quarterly* 23 (1979): 360–89; William Baugh and Michael Squires, "Arms Transfers and the Onset of War Part I: Scalogram Analysis of Transfer Patterns," *International Interactions* 10 (1983): 39–63; Rattinger, "From War to War to War," 1976.

24. Samuel Huntington, *Political Order in Changing Societies* (New Haven: Yale University, 1969); Geoffrey Kemp with Stephen Miller, "Arms Transfers Phenomenon," in *Arms Transfers and American Foreign Policy* ed. Andrew Pierre (New York: New York University, 1979). See also Talukder Maniruzzaman, "Arms Transfers, Military Coups, and Military Rule in Developing States," *Journal of Conflict Resolution* 36 (1992): 733–55.

25. See Ted Gurr, *Why Men Rebel* (Princeton: Princeton University Press, 1971); Douglas Hibbs, *Mass Political Violence: A Cross-National Causal Analysis* (New York: Wiley-Interscience, 1973); and R. J. Rummel, *Understanding Conflict and War* (Beverly Hills: Sage, 1975).

26. J. Bowyer Bell, "Arms Transfers, Conflict, and Violence at the Substate Level," in *Arms Transfers to the Third World,* ed. Ra'anan, Pfaltzgraff and Kemp, 1978, 309–23; Pierre, *Arms Transfers and American Foreign Policy,* 1979; Lars Schoultz, *National Security and United States Policy Toward Latin America* (Princeton, NJ: Princeton University Press, 1988); and Miles Wolpin, *Militarization, Internal Repression, and Social Welfare in the Third World* (New York: St. Martin's Press, 1986).

27. See Philippe Schmitter, "Foreign Military Assistance, National Military Spending and Military Role in Latin America," in *Military Role in Latin America,* ed. Philippe

Schmitter (Beverly Hills: Sage, 1973); Debbie Gerner, "A Statistical Study of Arms Transfers and Domestic Conflict in 57 African and West Asian Nations, 1963–1978" (Ph.D. diss., Northwestern University, 1982); James Dolian, "Military Coups and the Allocation of National Resources: An Examination of 34 Sub-Saharan African Nations" (Ph.D. diss., Northwestern University, 1973); William Avery, "Domestic Influences on Latin American Importation of U.S. Armaments," *International Studies Quarterly* 22 (1978): 121–42.

28. Ohlson, "Third World Arms Exporters," 1982; SIPRI, *The Arms Trade With the Third World*, 1971, 73.

29. Baugh and Squires, "Arms Transfers and the Onset of War Part II," 1983.

30. Baugh and Squires, "Arms Transfers and the Onset of War Part I," 1983.

31. Baugh and Squires, "Arms Transfers and the Onset of War Part I," 1983, 40.

32. Pearson, "An Analysis of the Linkage between Arms Transfers and Subsequent Military Intervention," 1981.

33. See also Donald Snow, *Distant Thunder: Patterns of Conflict in the Developing World*, 2d ed. (Armonk, NY: M.E. Sharpe, 1997).

34. Debbie Gerner, "Arms Transfers to the Third World: Research on Patterns, Causes and Effects," *International Interactions* 10 (1983): 5–37; Pearson, "An Analysis of the Linkage between Arms Transfers and Subsequent Military Intervention," 1981.

35. Baugh and Squires, "Arms Transfers and the Onset of War Part I," 1983.

36. Robert Jervis, "Cooperation Under the Security Dilemma," *World Politics* 30 (1978): 186–214.

37. This is a very slight modification of the Baugh and Squires presentation of this model. They use the term "Action-Reaction Process" in lieu of security dilemma. Baugh and Squires, "Arms Transfers and the Onset of War Part I," 1983, 40.

38. See especially Foster, "New Conventional Weapons Technologies," 1978.

39. Baugh and Squires, "Arms Transfers and the Onset of War Part I," 1983, 40.

40. Pearson, Brzoska, and Crantz, "The Effect of Arms Transfers on Wars and Peace Negotiations," 1992.

41. Baugh and Squires' model is: Arms Transfers→Military Capability→Victory (War Termination). The major difference between their model and the one used here is that they are concerned only with war termination, whereas this study is concerned with multiple indicators of war outcomes; i.e., winners, deaths, and length of war. Baugh and Squires, "Arms Transfers and the Onset of War Part I," 1983, 40.

42. Baugh and Squires, "Arms Transfers and the Onset of War Part II," 1983; Kinsella, "Conflict in Context," 1994; Kinsella and Tillema, "Arms and Aggression in the Middle East," 1995; and Sherwin, "Controlling Instability and Conflict through Arms Transfers," 1983.

43. Baugh and Squires, "Arms Transfers and the Onset of War Part II," 1983, attempted a similar evaluation using amount of conflict under way (lagged five years) and five-

year moving averages of weapons sales. Their use of a 10-year span to cover the effects of "slow replenishments for stocks drawn down by war" seem rather lengthy and thus shorter three- and five-year spans form the mainstay of this model.

44. See George Bohrnstedt and David Knoke, *Statistics for Social Data Analysis*, 3d ed. (Itasca, IL: F.E. Peacock, 1994); David Cox, *Analysis of Binary Data*, 2d ed. (New York: Chapman and Hall, 1989); Douglas Montgomery and Elizabeth Peck, *Introduction to Linear Regression Analysis* (New York: Wiley, 1982).

45. As there are no significance tests for pseudo-R^2, assessing results by this method alone would be inappropriate, Bohrnstedt and Knoke, *Statistics for Social Data Analysis*, 1994.

46. For an explanation of OLS, see Bohenstedt and Knoke, *Statistics for Social Data Analysis*, 1994.

47. See Melvin Small and J. David Singer, *Resort to Arms: International and Civil Wars, 1816–1980* (Beverly Hills: Sage, 1982); J. David Singer and Melvin Small, eds., *The Wages of War: 1816–1965, A Statistical Handbook* (New York: John Wiley and Sons, 1972); J. David Singer and Paul Diehl, eds., *Measuring the Correlates of War* (Ann Arbor: University of Michigan Press, 1990).

48. For a prominent criticism of the COW definitions, see Raymond Duvall, "An Appraisal of the Methodological and Statistical Procedures of the Correlates of War Project," in *Quantitative International Politics: An Appraisal*, ed. Francis Hoole and Dina Zinnes (New York: Praeger, 1976), 67–98. For alternatives, see Harvey Starr, "Revolution and War: Rethinking the Linkage Between Internal and External Conflict," *Political Research Quarterly* 47 (1994): 481–507; Harvey Starr and Benjamin Most, "Diffusion, Reinforcement, Geopolitics, and the Spread of War," *American Political Science Review* 74 (1980): 609–36; Steven Majeski and David Jones, "Arms Race Modeling: Causality Analysis and Model Specification," *Journal of Conflict Resolution* 25 (1981): 259–88; Guy Arnold, *Wars in the Third World Since 1945*, London: Cassell, 1995; Robert Harkavy and Stephanie Neuman, eds., *The Lessons of Recent Wars in the Third World, Volume I: Approaches and Case Studies* (Lexington, MA: Lexington Books, 1985); Robert Harkavy and Stephanie Neuman, eds., *The Lessons of Recent Wars in the Third World, Volume II: Comparative Dimensions* (Lexington, MA: Lexington Books, 1985); Istvan Kende, "Twenty-Five Years of Local Wars," *Journal of Peace Research* 8 (1971): 5–27; Istvan Kende, "Wars of Ten Years (1967–1976)," *Journal of Peace Research* 15 (1978): 227–41; Wright, *A Study of War*, 1942; Lewis Richardson, *Statistics of Deadly Quarrels* (Chicago: University of Chicago Press, 1960); and William Eckhardt and Edward Azar, "Major World Conflicts and Interventions, 1945–1975," *International Interactions* 5 (1978): 75–109, among others.

49. See SIPRI, *SIPRI Yearbook 1996, Armaments, Disarmament and International Security* (Oxford: Oxford University Press, 1996), 532–33; Edward Laurance and Joyce Mullen, "Assessing and Analyzing International Arms Trade Data," in *Marketing Security Assistance: New Perspectives on Arms Sales*, ed. David Louscher and Michael Salamone (Lexington, MA.: Lexington Books, 1987), 79–98; Edward Laurance and Ronald Sherwin, "Understanding Arms Transfers through Data Analysis," in *Arms Transfers to the Third World*, ed. Ra'anan, Pfaltzgraff, and Kemp, 1978, 87–105.

50. See Michael Brzoska, "The SIPRI Price System," in SIPRI, *SIPRI Yearbook 1987: World Armaments and Disarmament* (Oxford: Oxford University Press, 1987); and E. Sköns, "Sources and Methods," in SIPRI, *SIPRI Yearbook 1992: World Armaments and Disarmament* (Oxford: Oxford University Press, 1992).

51. Baugh and Squires, "Arms Transfers and the Onset of War Part II," 1983.

52. Undoubtedly, one could experiment with different estimation techniques. This is an area of potentially fruitful future research.

53. Baugh and Squires, "Arms Transfers and the Onset of War Part II," 1983.

54. For studies of the effectiveness of embargoes and economic sanctions, see Carter 1988; Doxey 1987; and Hufbauer and Schott 1985.

55. The fact that the United States is the unique supplier is no surprise here. The U.S. government has long maintained policies that have led to its classification as a "restrictive" supplier, according to SIPRI terminology; i.e., that foreign policy considerations, such as whether a state is engaged in warfare, are taken into account in arms transfer decision making sometimes at the expense of economic imperatives. For more on this pattern of supplier relationships, see Sanjian, "Great Power Arms Transfers," 1991.

56. It should be noted that these studies are based on different data, variables, and methods from Pearson, "An Analysis of the Linkage between Arms Transfers and Subsequent Military Intervention," 1981, and were *not* an attempt to replicate or *directly* test his intervention propositions.

57. All suppliers, except Czechoslovakia and the subsequent Czech and Slovak Republics, participated in at least one interstate or civil conflict during the years for which we have arms transfer data for these individual countries. Since the latter data varies widely given the differences in arms transfer histories of the states involved (e.g., because states such as North Korea have less than a 50-year history of selling weapons, the four major suppliers account for about 54 percent of the retained cases), supplier war involvement retained in the data set is 17 less than the actual post–World War II war involvement of the states examined here.

58. A contemporaneous model without the modification described in the text, i.e., with *all* arms transfer values in years in which wars broke out, produced the following results:

Model	−2 log likelihood	Goodness of Fit	Chi-Square	Predicted (Improvement)	Pseudo-R^2	Significance of B (sign)
Contemporaneous	758.347	689.160	34.581	74.85% (1.17%)	.02	.000 (+)

59. In these tests, data concerning wars that were ongoing at the time of the release of the last COW update (early 1993) were omitted because the status of much of the information of interest was necessarily unavailable.

60. Figure 3.5 was obtained by using the data of 52 cases of wars completed between 1950 and 1992 described earlier in the text. However, due to the extremely high casualties of war suffered by Iran and Iraq in their conflict (1.25 million combined

battle-related deaths), these cases proved to be extremely influential statistical outliers (e.g., Cook's D = 1.48 and Leverage value = 0.107 for Iran in the MAV3 model) and were omitted from the data plot and analysis provided in Table E in Appendix 3.

61. Hartung, *And Weapons for All,* 1994; Wolpin, *America Insecure,* 1991.

62. "Memorandum for Correspondents," 1997.

63. "Memorandum for Correspondents," 1997.

3

Appendix 3
More Details on Supplier
and Recipient Relations

**Table A: Country Statistics of Three-Year Prior Moving Averages
of War Termination Predicting Arms Transfers, 1950–1995**

Country	B	Standard Error	T-score	Significance of T
China	−168.215	124.992	−1.346	0.1859
Czechoslovakia	−140.967	85.102	−1.656	0.1046
France	−24.064	228.135	−0.105	0.9165
India	−11.659	5.437	−2.144	0.0477
Israel	7.083	27.589	.257	0.7989
Korea, North	26.225	19.740	1.329	0.2026
USSR/Russia	−1304.745	919.006	−1.420	0.1626
United Kingdom	60.326	108.621	0.555	0.5814
USA	723.794	632.252	1.145	0.2583

Table B: Country Statistics of Five-Year Prior Moving Averages of War Termination Predicting Arms Transfers, 1950–1995

Country	B	Standard Error	T-score	Significance of T
China	−134.791	204.118	−0.660	0.5128
Czechoslovakia	−333.878	128.233	−2.604	0.0124
France	−55.656	357.939	−0.155	0.8771
India	−1.320	10.168	−0.130	0.8983
Israel	64.439	41.517	1.552	0.1299
Korea, North	45.932	35.895	1.280	0.2189
USSR/Russia	−3066.040	1401.405	−2.188	0.0339
United Kingdom	101.013	170.368	0.593	0.5562
USA	1881.714	966.598	1.947	0.0578

Table C: Pooled-Time Series of Minor and Major Powers on Three- and Five-Year Moving Averages of War Termination Predicting Arms Transfers, 1950–1995

Group	B	Standard Error	T-score	Significance of T
Minor Powers, 3-Year	−63.769	45.696	−1.395	0.1648
Major Powers, 3-Year	−136.172	422.872	−0.322	0.7478
Minor Powers, 5-Year	−122.210	73.614	−1.660	0.0989
Major Powers, 5-Year	−284.742	663.431	−0.429	0.6683

Table D: Estimation of Weapons Imports on War Length by Ordinary Least Squares

Predictor	B	Standard Error	T-score	Significance of T
Contemporaneous	−0.007	0.012	−0.560	0.578
LAG1	0.003	0.007	0.425	0.673
LAG3	−3.615	0.006	−0.056	0.959
LAG5	−4.972	0.007	−0.062	0.950
MAV3	0.005	0.008	0.685	0.496
MAV5	0.002	0.008	0.271	0.787

**Table E: Ordinary Least Squares Results of Tests
of Propositions 7 and 8, Arms Transfer Effects on Battle Deaths**

Predictor	R^2	F-Score	Significance of F
Contemporaneous	0.002	0.10	0.75
LAG1	0.000	0.04	0.84
LAG3	.0005	0.27	0.60
LAG5	0.035	1.70	0.19
MAV3	0.016	0.80	0.37
MAV5	0.023	1.11	0.29

4

The Effects of Arms Transfers on War Involvement and Outcomes: Using Military Science to Test Basic Propositions Empirically

If the local situation appears to be in military imbalance, the side that believes itself stronger may be tempted to strike; if it does so, and if its assessment has been accurate, it is then likely that it will quickly overcome the victim (and violence will be ended). U.S. .... policy has supplied arms to avoid such local imbalances. Rationally, the sides are supposed to be mutually deterred from starting anything. But if hostilities nevertheless ensue they might well be more destructive and dangerous, and speedy termination might be much harder to achieve.

Lincoln Bloomfield and Amelia Leiss, *Controlling Small Wars*[1]

In any kind of regime the people who comprise the decision-making body are made of flesh and blood. Nothing would be more preposterous than to think that, just because some people wield power, they act like calculating machines that are unswayed by passions. In fact, they are no more rational than the rest of us—and indeed, since power presumably means that they are less subject to constraint, they may be less. A person whose life is governed solely by rational considerations pertaining to utility is, in any case, an inhuman monster. Now most decision-makers are not monsters; whereas those who are, such as Adolf Hitler or the Ugandan dictator Idi Amin, can scarcely be described as rational.

Martin van Creveld, *On Future War*[2]

Statement of the Problem

As noted in the statements by Bloomfield and Leiss and by Creveld, assertions concerning the rational calculation of military balances are prevalent in international affairs. In this regard, assertions are that it is not the absolute size of a state's military establishment that produces an inclination to conflict, but rather its size in relation to potential enemies. Further, many scholars have extrapolated, indicating that the spread of conventional weaponry has important effects on either the calculation of military balances or the correlation of forces in actuality.[3] Indeed, the statement quoted in the previous chapter from the U.S. government's notification of the sale of Harpoon antiship missiles to Egypt forthrightly proclaimed to Congress that the sales would not alter regional military balances and therefore cause instability in the Middle East. What is absent in this statement and elsewhere, however, is an appreciation of the precise mechanism of how conventional weapons affect the military balance between rival states or an explanation of how such calculations are made.

Arms sales are said to affect military balances in one of two ways, as noted by Bloomfield and Leiss. First, they may "stabilize" relations (promote peace between two states) by making adversaries more militarily equal. The alternative is also true: They may "destabilize" relations and cause war between states because they make adversaries militarily unequal. On the other hand, others posit that the likelihood of peace is *enhanced* when one state is clearly more powerful than another (one side has a clear preponderance of power). Thus, for these theorists, weapons transfers that preserve the power *imbalance* promote peace. Yet another theory contends that arms transfers have direct effects on war involvement; i.e., separate from any influence they may have on the military balance. These arguments assert that the impact of weaponry heightens leaders' awareness of military options to foreign policy problems. Because they desire some return on the heavy investment that they have made in acquiring military capability, they are more likely to adopt a militant foreign policy culminating in an increased probability that they will engage in war. However, categorical statements regarding the effects of weapons transfers or sales have garnered few in-depth studies, and little systematic evidence exists concerning the actual effects of arms transfers either on strategic assessment and then war, or directly on war itself.

In this chapter, we examine this phenomenon. In the first section, a review of several pertinent literatures concerning the importance of the military balance, the role of arms transfers on it, and the direct role that arms transfers may play on decisions regarding war is provided. The second section offers a research design, methods, and hypotheses based on the theoretical contentions. The third and fourth sections offer the models, data descriptions, and analysis of the impact of arms transfers on war outbreaks and outcomes, respectively. Finally, the concluding section summarizes the findings of the chapter and discusses the implications of this research.

Literature Review

The International Conflict Literature

Two distinct literatures contribute to a theoretical framework concerning the effect of the arms trade on war initiation and outcomes: (1) the international relations subfield of international conflict; and (2) the "arms trade literature" often associated with political economy. First, the international conflict literature posits competing theories concerning the role of militarized growth and the outbreaks and outcomes of war. Much of this literature is encapsulated in the debate between Michael Wallace, who found that arms races and military buildups lead to participant involvement in war, and Paul Diehl.[4] Diehl challenged Wallace's original findings concerning great power arms races and wars—and based on different data and operationalization of variables found that military buildups do not serve as an indicator of war.[5] Despite the disagreement in the literature (most of which is focused on great powers—who are capable of indigenous production of weapons —and great power interstate wars), weapons are agreed to be a necessary condition for modern warfare. However, the role of arms transfers in "causing" war is underexamined in this literature.

The international conflict literature also contains insights concerning the role of the "military balance" on war. These theories concern the effect of balance of power, power preponderance, and war.[6] Some scholars find that balance of power maintains peace through deterrence because neither side sees an "opportunity" to strike the other due to an advantage in power.[7] Others find that a clear imbalance preserves peace because power is concentrated so that one side is clearly far ahead of all challengers, and deterrence is provided because the weaker states do not challenge the strong.[8]

Additional insight into the puzzle concerning war involvement from the international conflict literature comes from Bueno de Mesquita and from Bueno de Mesquita and Lalman.[9] These studies provide the basis for a theory of war involvement at the level of the individual decision maker. They assert that leaders must make a complex calculation to determine the expected utility—all the probabilities and magnitudes of costs and benefits—for war with a given adversary. In no small part, this calculation is based upon another complex calculus: the military balance in terms of capabilities and intensity of preferences. Despite this important insight, Bueno de Mesquita does not assess the effect of weapons transfers on the calculus of the military balance in his expected utility model.[10] Nor does he assess the effects of arms transfers on the outcomes of wars.

Finally, T.V. Paul asserts that the acquisition of advanced weapons may give one side a temporary advantage in military power.[11] This advantage, when combined with other major variables such as particular type of military strategy and great power allies, may prompt decision makers to launch wars of aggression. This important work, while not focusing specifically or primarily on the role of arms transfers and not purporting to be a general theory,[12] provides a commendable effort to include (some) arms transfers in the study of (some) wars. This research, however, seeks more general findings than are available elsewhere.

While the conflict literature concerns itself with state war involvement, it is less developed when considering the outcomes of wars, saying little in a direct sense about the effect of arms transfers on war outcomes or even on whether arms transfers increase the recipient state's likelihood of winning.[13] The most prominent work in this area, by Allan Stam, asserts that total military personnel appears to have a major positive effect on war duration.[14] In other words, conflict between states with larger militaries may lead to longer wars. However, this ignores the role of technology, and therefore a primary impetus behind importing weapons, on war. More importantly, perhaps, he finds that the more out of balance the opponents' capabilities, the faster the war progresses. Unfortunately, his operationalization—functionally the same as Bueno de Mesquita's—says nothing about the effects of arms transfers. Thus, from his findings we know little about the effect of arms transfers, when considered independently from macroindices used as proxies for military buildups, and their effect on war outcomes.

Part of the explanation for the above lack of concern for arms transfers probably lies in the observation that international relations scholars have had trouble in measuring the "military" aspect of military balances. While the estimation of such balances are at least as old as the Bible, in the post–Second World War period approaches to these measurements have proliferated. Morganthau and many other realists urged the use of macrolevel indicators such as gross national product or Correlates of War (COW) power indices,[15] of state power in order to reflect the capabilities of states at total war.[16] Common to the approaches above is the failure to include arms transfers as an indicator leading to war involvement or affecting war outcomes. For the most part, these researchers have, curiously enough, ignored the role of weapons and weapons sales in their estimations of military balances. This is unfortunate because the use of aggregate, macrolevel data such as the COW project's indices fails to address the consequences that arms transfers may have on the military balance between competing states. For example, while the procurement of technologically advanced weapons might well have an important effect when one state confronts another state that perhaps has no adequate defense against the new and advanced weapon system, the effects of these acquisitions would not show up in macro measures.[17] Even the use of "military spending per soldier," as used by Huth, Bennett, and Gelpi as well as by Bennett and Stam, is insufficient to account for the "quality of military forces" when used alone.[18] Simply put, in a situation where the arms-producing state is willing to make financial arrangements—either outright arms transfers for free or through extensive offset arrangements—defense spending cannot capture the effect of greater technology on the battlefield (nor on military assessments of force balances) because it will not reflect the true cost of the weapons. Finally, it is doubtful that decision makers—those faced with the "complex calculus" of deciding whether to go to war, according to Bueno de Mesquita—would seriously consider an opponent's total population (for example) rather than its military might measured directly in terms of quantity and quality of its armed forces.

The Arms Transfers Literature

Within another, and largely separate, literature—that concerning arms transfers directly—there is a wealth of information concerning the numbers and types of weapons ordered and delivered by arms importers and exporters on a state-by-state level.[19] However, this literature has largely ignored the military effects of such transfers because they are "difficult to assess," because not all weapons bought and sold in the global marketplace are used in war.[20] They focus instead on the economic impacts of the weapons trade on suppliers and recipients. Others groups of analysts, especially those involved in the study of defense policy and the professional analysis of military science, do account for numbers and types of weapons in a state's inventory and use a "bean counting" technique.[21] In this method, analysts compute the correlation of forces based on numbers of weapon systems; i.e., country A has 250 main battle tanks versus country B's 275, 145 combat aircraft to 100, etc. The weaknesses of this technique are well described in Sherwin and Laurance.[22] These scholars instead rely on weapons capability scores.[23] Unfortunately, their technique was never validated by actual research or empirical examination. When combined with models that purport to represent the interaction of forces in combat, weapons capability scores provide information concerning attrition (casualties), front displacement (territorial loss), duration of war, and war winners. However, to date, these models have not been used to analyze the effect of arms transfers on war, as will be discussed at length below.

Other studies use historical analysis of individual states, such as works by Kinsella and by Kinsella and Tillema on the effects of the superpower rivalry as exhibited through arms transfer policies to the Middle East, in order to posit general conclusions.[24] While these methods are at times appropriate for the examination of empirical questions focused on specific cases, they leave open the question of how applicable their findings are across cases. Others have studied the effects of weapons transfers on regional relations; while yet others provide qualitative insights on a more extensive range of cases by using historical analysis.[25]

The most extensive and general of the above studies are by Brzoska and Pearson and by Pearson, Brzoska, and Crantz.[26] These studies assert that several basic conclusions can be derived from studies concerning the effect of arms transfers. Foremost among these are assertions that "arms deliveries are a factor in decisions to go to war because of considerations of military superiority [and] perceptions of changes in the balance of power."[27] Thus, according to these researchers, states are more likely to go to war after receiving arms transfers that change the perception of the balance of power. Furthermore, "arms transfers generally prolonged and escalated wars, resulting in more suffering and destruction."[28] Because of the influx of weapons into an ongoing conflict, the flagging strength of the belligerents is purportedly replenished, allowing them to continue with the war. Furthermore, according to Brzoska and Pearson, analysis of arms transfer flows may provide hints as to who will win the war. They say, "It is best not to deliver any weapons during conflict, unless a supplier favors one side to win."[29] While these important studies provide insightful analysis of the past influence of arms transfers on war, in effect the predictive power of such efforts is hindered by the "too few cases, too many variables" problem of attribution of causal effect. Furthermore, the authors are unclear in describing their underlying method for assessing such important factors as "pre-war [capability] advantage" and how the term *asymmetric*, defined to indicate that the "named party received substantially more arms," is weighted (i.e., whether the quality as well as quantity of weapons was assessed).

A second, and separate, theoretical relationship is posited by the Stockholm International Peace Research Institute.[30] This research indicates that the sale of weapons has an effect on the decision to engage in war separate from any influence that weapons imports have on the perceived military balance. By devoting resources to the importation of weapons (especially those considered "high tech" or in major weapons categories), leaders become more attuned to military options when foreign policy decisions are required. Because of this heightened attenuation, weapons are said to have a powerful and *direct* influence on decisions to engage in military activity and especially war.

Unfortunately, researchers in the area of arms transfers seem wont to declare, as does Catrina, that "the question of what would have happened if an attack had occurred in the absence of a strengthening of the recipient state's military capabilities" is worth asking, but apparently only in a rhetorical sense.[31] This question, though, is of primary importance if we are to make assessments of the impact of arms transfers. Simply put, if arms transfers *do* serve to increase the probability of war outbreak, it is because they affect leaders' calculations of the "new" military balance. This balance is, in turn, nothing more than a projection of what would happen if war broke out (see Figure 4.1).

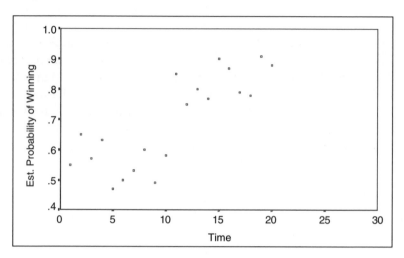

**Figure 4.1: Reassessment of Probability of Victory
in War after an Arms Transfer**

To show that arms transfers have an impact, then, we must show that there is a difference between what "would have happened" had war broken out in the absence of the new weapons (represented by the data points to the left of Time = 10 in Figure 4.1) and what is projected to happen with their presence (represented by the data points to the right of Time = 10 in Figure 4.1). The hypotheses and models below are designed to allow the empirical examination of this.

Research Design on Arms Transfers
and War Involvement: Relationships,
Major Hypotheses, and Methods

The first hypothesis to be considered provides further examination of Pearson, Brzoska, and Crantz's important assertion that "arms deliveries are a factor in decisions to go to war because of considerations of military superiority [and] perceptions of changes in the balance of power." Not all arms transfers, however, will lead to an increased likelihood of war according to this logic. Rather, only those transfers that alter the military balance are more likely to lead to war. Thus, military balances can be seen as an intervening variable in the following relationship:

$AT \rightarrow \Delta MB_{A/B} \rightarrow \rho War;$ where

AT = Arms transfers;

$\Delta MB_{A/B}$ = Change in Military Balance between states A and B;

ρWar = Probability of War

(a)

A further caveat is in order due to the limited power projection capability of most states in the international system at the present time and throughout history. Because of the limitations of states when projecting power beyond their own immediate borders, only arms transfers that upset the military balance between bordering states will be considered.[32]

To test the relationship above, we must first assess the assumption that arms transfers change the military balance between states. It is possible that weapons imported may not, when the combined military capabilities of a state are aggregated, change the military balance to any statistical or substantively significant degree. Our first hypothesis, then, accounts for this issue.

- *Hypothesis 4-1:*
  Arms transfers make a significant impact on the military balance between rival states.
  If this hypothesis is true, then the following hypotheses can be investigated:

- *Hypothesis 4-2:*
  Given a contiguous dyad, arms transfers that upset the military balance increase the probability of war involvement of the dyad members.
  Secondly, to take into account the discrepancies in the theories, we posit that more evenly balanced dyads will be more likely to fight and that dyads where relative power is asymmetric, peace (or lessened probability of war) is more likely.

- ■ *Hypothesis 4-3:*
  Given a contiguous dyad, arms transfers that create conditions of greater parity in military capabilities increase the probability of war involvement of the dyad members.

If Hypothesis 4-1 proves false, however, and there is no significant change in military capabilities wrought by arms transfers, then the examination of Hypotheses 4-2 and 4-3 are rendered logically irrelevant, and we must examine an alternative hypothesis. These hypotheses keep the spirit of Pearson, Brzoska, and Crantz, while discounting the effect on the military balance. The final hypothesis is designed to reflect the SIPRI assertion that arms transfers increase the likelihood of war involvement because they heighten leaders' awareness of military options. Thus, it considers a direct relationship between arms transfers and war involvement—one that circumvents the argument that they affect such a relationship via changes in the military balance. The following model expresses the relationship here:

$$AT | Ratio \rightarrow pWar \qquad\qquad (b)$$

where arms transfers directly affect the probability that a state is involved in interstate war with its neighbors, while holding the ratio of the military capabilities of the respective sides (ratio) constant.

- ■ *Hypothesis 4-4: Given a contiguous dyad and an established ratio of military capability between the states of the dyad, arms transfers increase the probability of war involvement between the dyad members.*

Methods

In order to test the Hypotheses 4-1 above, it is necessary to assess whether there is an appreciable effect that the transfer or sale of weapons have on the military (im)balance between contiguous states. To do this, we perform a difference in means test between the pairs (i.e., with and without arms transfers) of military balance ratios (the calculation of which is described below). Regardless of the results of this test, the method for assessing Hypotheses 4-2, 4-3, and 4-4 is logistic regression. Logistic regression[33] is a statistical technique in which the constructed model predicts certain outcomes, which are then compared to a null hypothesis, which states that the phenomenon under examination (the outbreak of war, in this study) never occurs. From this, the researcher can observe the value of the model by aligning predicted cases with actual cases (i.e., Did the model predict war in this case?), where more powerful models are better at predicting cases correctly.

In these tests, we analyze whether such changes affected the likelihood of war involvement in two groups: a test group of states involved in interstate wars from 1969 to 1990,[34] and a control group of matched states not involved in interstate wars. For the data, there are 11 wars in which the belligerents shared borders (see Appendix 4). These dyads make up our test group. For the control group, 10 border-sharing dyads were chosen. Within these match dyads, at least one state was matched with a counterpart in the test group based on three criteria.[35] First, a state was chosen within the first year of war involvement by the state in the test set. Second, every attempt was made to match war set states with control set states from the same geographic region. Third, control set matches match the test state within ±15 percent on energy consumption in order to account for similarities in developmental stage, military spending (similarities in military policies and emphasis) and GNP (similarities in size of the economy). When this condition could not be satisfied within the region, out-of-region matches were made with countries in regions of similar terrain, and with states of similar energy consumption, military spending, and GNP.

Before moving on to the models, an explanation of how we will measure military capabilities and their ratios is necessary. For this study, the military capabilities of states are measured in terms of an aggregate Weighted Effectiveness Index (WEI), commonly used in military models (and described fully below) and by the Predicted Outcome Indicators (POI), which is a term used here for the outputs of force attrition models used by military scientists in force planning (also discussed below). Because indicators such as the WEI and POI are virtually unknown in the political science literature outside of the subfield of strategic studies, a short review of their application elsewhere is warranted.

WEIs and POIs and Military Science

Military science and strategic studies have long been concerned with two of the primary issues related to the study of war: the relative measure of combatants' (or potential combatants') military capability, and the precise mechanics of how the capabilities of each side in combat interact to determine outcomes.[36] By using the measurements and models developed in this literature, we can assess more accurately the impact of weapons on the likelihood of a state's war involvement and the outcomes of the war than is possible using other techniques (such as "bean counting" or using a state's military spending or GNP as an indicator).

The most widely used means of measuring the military capabilities ratio between warring sides within the military modeling literature is the WEI.[37] It combines the quantitative aspects of a country's individual military equipment with a qualitatively derived measure of how effective this equipment is per weapon category (e.g., tanks, artillery, and helicopters are deemed as varying in degree of lethality as a group), thus allowing a better assessment of force ratios than simply counting weapons such as tanks and aircraft.[38] WEI scores are absent for fixed-wing aircraft because they are often considered as a separate part of the battle in military models.[39] We create approximations for these weapons by using insights from within the military science literature. We construct approximate aircraft WEI values based on Dupuy's data showing that a fighter-bomber (MIG 23) of average quality is roughly twice as effective as a T-55 main battle tank, or a 1.5 weighting.[40] His methodology also establishes that fixed-wing and rotary aircraft can be considered similar vehicles because of their high mobility. Thus, we use the terrain weights for attack helicopters provided by Raymond because at ground attack speeds and altitudes, fixed-wing aircraft are roughly equivalent to helicopters (greater payload, but less accuracy).[41]

A final note: More modern formulations of the WEI allow for the dynamic change in effectiveness of weapons due to such factors as differences in terrain, weather, and night.[42] Table 4.1 provides an illustrative example of how the calculation of aggregate WEIs are performed.[43]

Two other aspects of a country's military capability must be taken into consideration: (1) military quality (e.g., How well trained are they?); and (2) diffraction due to number of bordering states (i.e., If country A has three bordering states, it cannot commit all of its forces against opponent B, because that leaves it unacceptably vulnerable to opponents C and D). Thus, the aggregate WEI for country A is calculated as follows:

$$WEI_A = \frac{Aggregate\ WEI \times Military\ Quality}{Number\ of\ Bordering\ States}$$

where

$$\text{Military Quality} = \frac{\text{Military Spending}}{\text{Number of Military Personnel}} \quad 44$$

From these values we calculate the weighted military balance by dividing the war initiator's WEI by the defender's WEI and deriving a ratio. By doing this at two points in time, before and after weapons sales, we can test the difference of means model and Hypothesis 4-1. For the purposes of evaluation, we count all weapons sales of major combat vehicles over the three-year period immediately preceding the onset of war (or the hypothetical onset, for the control cases).

Table 4.1: Method of Calculating Aggregate Weighted Effectiveness Index

Weapon Type	*Quantity (a)*	*Weapon Value (b)*	*Terrain Value* (c)*	*Weapon WEI (a × b × c)*	
T-72 (tank)	95	.75	–4.20	299.25	(line 1)
BMP-1 (armored personnel carrier)	210	1.00	6.77	1421.70	(line 2)
MI-8 (helicopter)	45	.75	16.04	541.35	(line 3)
D-30 (towed artillery)	20	.70	9.58	134.12	(line 4)
			Aggregate WEI	2396.42 (sum lines 1-4)	

*   Terrain values provided in this illustration are for mountainous conditions/daytime.

The second measurement of military capability ratio, the POI, is derived from using the WEI values in a force attrition model and measuring the change in projected war outcomes given the ratio and types of forces of the belligerents. Modeling military balances and the outcomes of potential and actual wars is an old, albeit imperfect, art. Perhaps the earliest, and certainly the most influential, of the military modelers was Frederick Lanchester.[45] Lanchester's equations, and military models in general, seek to capture the dynamic nature of warfare by representing the size and capabilities of opposing forces and the attrition rates of battle.[46] The application of Lanchester's equations to such wide-ranging conflicts as the Thirty Years War to futuristic scenarios in the Persian Gulf has met with mixed results.[47] Nonetheless, assessments of this type have for decades dominated the dynamic assessment of conventional (and air) balances and have been—and are still—used by such entities as the U.S. Army, the Joint Chiefs of Staff, the Office of Net Assessment (and other directorates) of the Office of the Secretary of Defense.[48] They are used by independent defense analysts and form the core of the force planning curricula of major universities and postgraduate service schools. Finally, they form the basis of theater-level combat modeling conducted under contract for the Pentagon and are central to the scholarly literature on conventional war gaming and simulation.[49]

Despite the disagreements over the historical accuracy of Lanchester models when it comes to "post-dicting" actual wars, military models retain a vital function within the realm of force planning.[50] According to one analyst, assessments of "conventional military balances ... dominate American defense planning."[51] They also play a role in war involvement decisions, that war "is a dispute about measurement [of relative strength]; peace on the other hand marks a rough agreement about this measurement."[52] Defense analysts' means of assessing military balances are of primary interest here, because by using their formulations the impact of additional weapons on military balances can be assessed. As noted by Quester, the estimation of combat capability is a problem at all levels of conflict—strategic, theater, operational, and tactical.[53] Furthermore, miscalculation is one of the fundamental causes of war; in other words, nations have fought wars based on the mistaken belief that they would emerge victorious. Finally, because these are force-on-force or force attrition models, they can be used to examine the effects of arms transfers on war involvement (as they allow for assessments of pre- and postarms transfer military balances) and outcomes (model outputs are typically given in terms of casualties and duration).

Force attrition models more dynamic than Lanchester's are of a type pioneered by Joshua Epstein and are the state of the art in military science (although many operations analysts continue to use Lanchester's lawlike formulations in combat models).[54] Epstein's models, unlike models based on Lanchester's equations, account for the ability of defenders to trade space for time, the need of defenders or attackers to lessen the attrition rate under which they fight, and the diminishing marginal returns that may occur in stepping up the tempo of an attack. In his "feedback" model, Epstein describes war as an uneven series of velocity curves that reflect the outcomes of combat through the interplay of two adaptive systems, "each searching for its equilibrium."[55] This interplay results in the territorial loss that occurs, the casualties suffered by each side and the duration of the conflict, thus providing a more dynamic and complete representation of war than do Lanchester variants (i.e., whereas Lanchester models give duration and casualties, Epstein gives duration, casualties, and front displacement).[56]

Based on force attrition models such as Epstein or Lanchester, the analyst can project the outcome of a potential conflict between any group of adversaries. Upon the basis of such projections, force planning decisions are made. Because the acquisition of arms (either indigenously or from external sources) alters the parameters of force planning models, these models can be used to assess whether transfers have tended historically toward military balance or imbalance. However, military models have not been used to gauge the effect that arms transfers would have on military balance. Nor have they been used for the sake of estimating the likelihood of war involvement due to arms transfers. Neither have they been applied to estimating the effect of arms transfers on war outcomes.

When we use the Epstein model to derive the POI, the technique described above for incorporating aircraft into the aggregate WEI values for a state will not be necessary because in this model, aircraft are not added to ground lethality, but rather "attrites" it (i.e., subtracts from it). Thus, for the POI model, aircraft are counted in raw numbers, and we can assume (as does Epstein) historical maintenance, loss, sortie, and kill per sortie rates to affect the respective airpower and ground lethality totals in the model. Of course, to test Hypothesis 4-1, the force attrition model was run twice. First, it was run with the additional military capability gained from the arms transfers included. Then it was run a second time omitting the additional military capability gained by arms transfers over a given period immediately preceding the war.

For example, Epstein calculates that an attacker with a strength of 330,000 (in aggregate WEI equivalent) would take 57 days to bring a war with a defender with a strength of 200,000 to an end (and will lose—the attacker's force strength would go practically to zero). This war will cause the attrition of 99 percent of the attacker's forces. During the hypothetical war, the attacker will occupy 76.6 km of the defender's territory. Let us assume that these are the results without (or prior to) the benefits of any arms transfers. In order to derive the POI, we would add in the final WEI strength attained by the attacker and defender through the importation of weapons (in this case, let's assume that only the attacker imports weapons). With this "new" calculation of force strengths, we can run the model again, and again obtain the estimated outcome of conflict in duration, casualties, front displacement, and winner/loser (by letting one side's strength go to zero) and compare them to the initial estimates (as in Table 4.2).

**Table 4.2: Example of Predicted Outcome Indicator in the Examination
of the Effect of Arms Transfers on War Outcomes**

	Duration	*Casualties*	*Front Displacement (kms)*	*Winner/Loser*
Initial Estimate (no AT)	20 days	99%	76.6	Defender/Attacker
Revised Estimate (with AT)	10 days	86%	108.8	Attacker/Defender

From these two estimates of POI, the difference of means tests are
calculated for Hypothesis 4-1. If there proves to be a significant effect by
arms transfers on the military balance exhibited by this test, we can
easily calculate the percentage of change made by arms transfers from
these figures. Thus, the gain in relative capability due to arms transfers
from the transaction posited above is, depending on how one wishes to
measure it, 50 percent duration, 13 percent casualties, 30.6 percent front
displacement, or 100 percent winner/loser. Measuring relative capability
gained through arms transfers in this way seems most sensible given
Bueno de Mesquita's assertion that decisions to go to war hinge, in part,
on the calculation of probable outcome; yet this has never been done in
the literature. Using multiple indicators (i.e., duration, casualties, front
displacement, winners/losers) is intuitive given the uncertainty as to
which of these factors may play the dominant role in such calculations
(i.e., would a leader go to war after receiving arms transfers if it were
estimated that the war would last half as long, or because it would result
in fewer losses, etc.?). Finally, the force planning models used here are
slight modifications of Epstein's dynamic model (see Appendix 4).[57]

Data Analysis on the Effect of Arms Transfers
on War Involvement: Models, Data, and Results

Model 1

In order to evaluate a model that eventually will measure the effects
of arms transfers on war, the issue of the alleged intervening
variable—military balance—must be first considered. We do this by a
difference of means test of the following model:

$$\frac{Milcap_a}{Milcap_b} \neq \frac{MilcapAT_a}{MilcapAT_b},$$

(Model 1)

where $Milcap_a / Milcap_b$ is the military capabilities ratio (measured in aggregate WEI scores) between the two states of the contiguous dyad without arms transfers, and $MilcapAT_a / MilcapAT_b$ is the ratio with the capability values of arms transfers added in. When the POI is used, we use the actual raw numbers in terms of duration, attrition, front displacement, and winners in order to ascertain whether arms transfers have a significant effect on the perceived outcome of a war; i.e., do arms transfers have a significant effect on the predicted duration of the war? The attrition suffered? Territory lost? Who wins?

Data

For the independent variable, WEI values will come from Raymond, who developed them for the use of the U.S. Army's Command and General Staff College.[58] Raymond's dynamic WEI formulations are based on the earlier work of the U.S. Army's Concepts and Analysis Division (which have been in use by NATO-country defense establishments for over 20 years).[59] POI measures are derived from the results of running the initial WEI data in a variant of Epstein's model. The makeup of the countries' armed forces will be taken from *The Military Balance,* published by the International Institute for Strategic Studies. While these data suffer from very real aggregation problems,[60] there are no superior open-source compilations that use a consistent coding methodology, have global coverage, and are available for extended time periods.[61]

For the military capabilities without arms transfers, we will use the total military capabilities from combining the WEI values with *The Military Balance* information as above, and then subtract the total WEI valuation of arms transfers for the recipient country over the three years prior to the outbreak of war. For the arms transfers data, we use the SIPRI dataset. Of the three institutions that publish arms transfer data sets of varying types, SIPRI, the U.S. Arms Control and Disarmament Agency (ACDA), and the United Nations, only SIPRI data can be used for the analysis in this chapter.[62] The SIPRI data set, already discussed in Chapters 2 and 3, is an open-source compilation of transfers of major weapon systems by year(s) of delivery. Of great importance for this study is the fact that SIPRI data are available in "raw form," which allows the researcher to reconstruct arms transfers by military item (rather than on a dollar or index basis) yearly from 1950 to the present. Using SIPRI data in its raw form (which provides both specific weapon type—e.g., M48A3—as well as the following weapon categories: tanks, artillery [including self-propelled, towed, mortars, and MRLS systems], fixed-wing aircraft and helicopters, and armored personnel carriers)[63] will also avoid one of the most controversial aspects of both SIPRI and ACDA data sets—derivations of approximate monetary or indexing the values of the weapons. In effect, we will take the numbers and types of weapons sold from SIPRI, and use the WEI methodology developed for use by the U.S. Army's war colleges and analysis sectors to construct an alternative index.

Results: Testing Hypothesis 4-1, Do Arms Transfers Have a Significant Effect on the Military Balance?

To test Hypothesis 4-1, we examine whether arms transfers significantly change the military balance for the historical data. If arms transfers do change the military balance, then we may proceed to the test of Hypotheses 4-2 and 4-3. If not, then we can rule out the intervening effects of arms transfers on the military balance and proceed to examine whether there is a more direct effect between arms transfer values and war (Hypothesis 4-4). Further, if arms transfers are found to not change the military balance, then we will already have increased the scientific body of knowledge in this area.

In order to determine the effect of arms imports on military force ratios, a difference in means test is conducted (see Table 4.3 for results). In this test, we examine whether there is a statistically significant difference in the average change in force ratios between attacker and defender of the test and control groups. The results, using the modified aggregate WEI values described above, indicate that the differences between pre- and post-arms transfer military balances are statistically indistinguishable. Likewise, when we look at the POI indicators, we find that in only one of the four indicators are the differences in model outcomes significant between with-transfer and without-transfer data runs (see Table 4.4).

**Table 4.3: Comparison of Means of Pre- and Posttransfer
Weighted Military Force Ratios (Aggregate WEI Values)**

Dyad	Year	Pretransfer Ratio	Posttransfer Ratio
India-Pakistan	1971	2.636171	2.777976
Vietnam-Cambodia	1975	1.659338	1.745594
Arab-Israel	1973	0.587445	0.697332
Iraq-Kuwait	1990	1.392313	0.974615
Israel-Syria	1982	13.215900	6.733835
Uganda-Tanzania	1978	1.033349	1.128337
Somalia-Ethiopia	1977	0.297777	0.230476
China-Vietnam	1985	3.600211	3.574402
Iraq-Iran	1980	0.759922	0.776302
China-Vietnam	1979	7.980367	7.776302
Israel-Egypt	1969	2.66016	1.783292
Thailand-Burma*	1971	2.033825	2.145454
Malaysia-Singapore*	1975	0.843090	1.064964
Afghanistan-Iran*	1973	1.808456	2.352552
Israel-Jordan*	1990	32.314490	30.396290
Libya-Tunisia*	1982	16.870610	16.873660
Gabon-Cameroon*	1978	1.445873	2.425616
Guinea-Sierra Leone*	1977	2.301056	2.333799
Bangladesh-Burma*	1979	2.005581	2.032785
United Arab Emirates-Oman*	1980	2.075162	2.372808
Bangladesh-Burma*	1985	2.167421	2.209803
	MEAN	4.747100	4.400200
	t-value		−1.050000
	2-tail significance		0.305000

*   indicates control dyads of matched pairs.

**Table 4.4: Difference in Means of Predicted Outcome Indicators
with and without Arms Transfer Values[64]**

Variables	Difference in Mean	t-value	2-tail significance
Duration (with AT)-Duration (without AT)	−7.2857	−2.13	0.046
Attrition (with AT)-Attrition (without AT)	−7.9048	−1.59	0.127
Front Displacement (with AT)-Front Displacement (without AT)	−35.6239	−1.63	0.119
Winner (with AT)-Winner (without AT)	0.1905	7.15	0.162

From the results of the WEI and POI means of evaluating the affect of arms transfers on the military balance, it appears that arms transfers largely *do not* affect the military balance at this level of aggregation.[65]

Thus, we will reject Hypothesis 4-1 (and also conclude that the study of Hypotheses 4-2 and 4-3 would be illogical), and move on to the investigation of Hypothesis 4-4.[66]

Model 2

To test Hypothesis 4-4, we create the simple model:

$$\frac{Milcap_{a(t)}}{Milcap_{b(t)}} + \sum_{i=3} AT \rightarrow War\ Involvement$$

(Model 2)

where $Milcap_a$ is the military capabilities of side A (attacker), and $Milcap_b$ is the military capabilities of side B (defender) at the first day of the first year of the war, and $\sum_{i=3} AT$ is the total amounts of arms imports (measured in WEI terms) over the preceding three-year period.

For this model, the dependent variable is a binary term for whether the states under consideration were involved in an interstate war in a given year (1 = war, 0 = no war).

Data

The compilation of interstate wars comes from the COW project. For the independent variables, the ratio of military capabilities will be measured as aggregate WEI values as described above. Arms transfer values will come from SIPRI and will be aggregated according to the WEI values assigned by Raymond for the prevailing terrain type within the theater of war (or hypothetical war).[67] Daylight operations are assumed, because most countries in our data have very limited capabilities for night fighting. When the POIs are used as the control variables, they are coded as follows. For the predicted duration variable, the number of predicted days of war is coded as negative for the war winners and positive for losers. This is because, besides the inherent value of shorter wars in terms of destructiveness to the winners' forces, winners will desire shorter wars so that other powers have less chance to intervene to save the opponent. The opposite is the case for the loser, due to the self-evident logic that a losing state would like to increase the amount of time that it could hold out. Predicted attrition is coded as the negative percentage of beginning total forces lost by the attacker and defender. The negative value is taken because higher attrition rates will be perceived as detrimental to the effort of either side. Predicted front displacement, or territory lost, is coded as the positive sum of front displacement (in kilometers) for the attacker, and the negative sum for the defender. This logic is self-evident. Note that even when the defender "wins" the simulated war, it must perceive that "losing" territory will be costly to regain either militarily or at the peace table. Finally, predicted winners is a dummy variable that represents the hypothetical outcome of the simulation if we let the either the attacker's ground force strength go below 10 percent, or the defender's go to zero (literally, the defenders will hold out "to the last man"). For this variable, a win is coded as a "+1" and a loss is coded as "–1."

Results: Testing Hypothesis 4-4, Do Arms Transfers Affect the Probability of War Involvement between Contiguous States Even if Such Transfers Have No Effect on the Military Balance?

In the investigation of whether arms transfers affect the likelihood of whether a state engages in war against a bordering state, Table 4.5 shows that there seems to be considerable evidence to support such a contention. In this table, $\sum_{i=3} AT$, the variable for the WEI amounts of weapons imported into a country in the three-year period prior to the outbreak of war (or hypothetical outbreak for the control group) is a strong and positive indicator of war involvement.

Table 4.5: The Effect of Arms Transfers on War Involvement When WEI Ratio Controlled

N	–2 Log Likelihood	Goodness of Fit	Model Chi-Square (significance)	
42	39.507	47.633	18.622 (.0001)	

Variable	B	Std. Error	Significance	R
$\sum_{i=3} AT$ (WEI)	0.0002	9.285	0.017	0.251
RATIO$_t$ (Aggregate WEI Values)	–0.0570	0.075	0.450	0.000
Constant	–0.8944	0.529	0.019	—

From this model we can also observe the reduction of errors when compared to a null model which predicts that war will always occur (and is thus correct 53.4 percent of the time). In Table 4.6 below, we see that Model 1 performs admirably in this task, providing a 27.5 percent improvement over the null model. Table 4.6 also indicates that the model only makes one false positive prediction for the control group, while wrongly predicting no war participation by the test group in seven cases.

Table 4.6: Classification Table for Arms Transfers Predicting War Involvement when WEI Ratio Controlled

Observed	Predicted		Percent Correct
	No war	*War*	
No war	19	1	95.00
War	7	15	68.18
		Overall	80.95

When we control for the alternative measure of military balance, the predicted outcome indicators (POIs), we find very similar results. Turning to Table 4.7, we again see that the $\sum_{i=3} AT$ variable for arms transfers prior to the initiation of war involvement is a significant, positive indicator of the states' war involvement. In addition, the predicted winners variable shows a significant, but negative relationship with war involvement. This is, however, a statistical anomaly. Close scrutiny of the data reveals that of the six cases in the test and control sets where the defender is predicted to win, only one case is located in the control set (Malaysia defeats the Singaporean attack). Thus, whereas in about half (5 out of 11) of the test set cases (where war involvement = 1) the defender wins (coded "–1"). In the control set (war involvement = 0) only 1 out of 10 results in defender wins (coded "–1"). This provides a negative relationship between predicted winners and war involvement, albeit one that may be spurious when considered in the light of this study.

Table 4.8 provides the classification table for the model above. There we find that the predictions of the model where POIs are controlled are very similar to that of the WEI ratio model, with two more cases of war wrongly predicted in the control set, and two additional cases correctly predicted in the test set (thus providing the same overall predictive power).

Table 4.7: The Effect of Arms Transfers on War Involvement
with POI Estimations Controlled

N	−2 Log Likelihood	Goodness of Fit	Model Chi-Square (significance)	
42	*35.461*	*35.255*	*22.668 (0.0004)*	
Variable	*B*	*Std. Error*	*Significance*	*R*
$\sum_{i=3} AT$ (WEI)	0.0002	9.947	0.014	0.260
Duration	0.0016	0.019	0.935	0.000
Attrition	0.0140	0.015	0.352	0.000
Front Displacement	−0.0006	0.003	0.849	0.000
Winners	−0.9743	0.500	0.051	−0.175
Constant	0.3051	1.135	0.788	—

Table 4.8: Classification Table for Arms Transfers Predicting
War Involvement with POI Estimations Controlled

Observed	Predicted		Percent Correct
	No war	*War*	
No war	17	3	85.00
War	5	17	77.27
		Overall	80.95

To summarize, in this section we found that when using WEI and POI formulations of military capabilities balances before and after arms transfers, there is—quite surprisingly given the assertions of prominent theories of the effects of arms transfers on war involvement—no significant difference between the two. Thus, we rejected Hypothesis 4-1 —that arms transfers change the military balance between states—and declined to investigate Hypotheses 4-2 and 4-3 because they are precluded logically by the rejection of 4-1. In the investigation of Hypothesis 4-4, we found that when controlling for the military balance at the time of the war (or hypothetical war) outbreak, the amount of arms transfers in the three-year preceding period was a statistically significant and positive predictor of war involvement. The resulting models (WEI and POI derivatives) were able to correctly predict 34 of the 42 cases in the data set, and were thus a 27.5 percent improvement over a naive model, which only predicted correctly about 53 percent of the time. It is worth emphasizing again at this point that it is quite possible that when better data become declassified or are created, this record could improve considerably. Yet despite the weaknesses of the current data, these tests indicate that it is quite possible that the international conflict literature has ignored an important variable in the study of war—the arms transfer.

Research Design on Arms Transfers
and War Outcomes: Major Hypothesis

Next, we move on to the empirical examination of the proposition that arms transfers prolong wars. For these tests, a war has already begun between two or more bordering states, and we seek to measure the effects of arms transfers to one or both sides on the outcomes of the war. The ensuing hypothesis shall be tested:

- *Hypothesis 4-5: Weapons transfers into a contiguous borders conflict dyad are associated with measurable, direct, and significant alterations of the outcomes (duration, winners, and casualties) of the war.*
  The null hypothesis is that arms transfers are not associated with a significant effect on outcomes of wars.

Methods

For this hypothesis, concerning impacts of arms transfers on war outcomes, two statistical techniques are used—logistic regression and ordinary least squares. As logistic regression was used above and ordinary least squares is commonplace in the disciplines of political science and international relations, neither necessitate further description here.

Data Analysis on the Effect of Arms Transfers
on War Outcomes: Models, Data, and Results

Model 3

Beginning with the logistic regression model, we examine the effect of arms transfers on whether a recipient state wins or loses a conflict. The theory indicates that arms transfers are beneficial in terms of helping the recipient state win the war. The model for this is:

$$\frac{Milcap_{a(t)}}{Milcap_{b(t)}} + \sum_{i=3} AT \rightarrow Winner$$

(Model 3)

In this model, war is ongoing and we measure whether the state is the winner in binary terms (1 = winner, 0 = loser). $Milcap_a/Milcap_b$ is again defined as the ratio of aggregate WEI forces of the opposing sides at the beginning of the war in one instance and in terms of the POI in the other (same codings as used above). Since we found above that arms transfers have no distinguishable influence on the military balance, we again operationalize them to indicate their separate effect. $\sum_{i=War} AT$ is the WEI values in arms transfers delivered to the recipient during the years of war. When necessary, the WEI values will be modified to reflect the portion of the year that has expired before the war begins; i.e., if a war begins in June 1987, the value for 1987 would be $WIE_{1987}/2$ because half of the year of 1987 had expired before the war began.[68]

Data

The data for military capabilities and arms transfers have been described above. The major differences in the data set used here are that we are only using the "war" or test cases of the data set above, and that the POI for predicted winner must be altered. For predicted winners, the coding was previously: "attacker wins (+1)/defender wins (–1)," so that we could control for the predicted victor in assessing a state's probability of *war involvement*. Since we are no longer interested in war involvement, but are instead investigating the question of whether arms transfers affect the likelihood of winning for a state given the *predicted outcome* of the war provided by a scientific model, we must recode this variable. This has been achieved by recoding predicted winners to a case-by-case dummy variable for the winner and loser of each war (e.g., for the India-Pakistan War in 1971, India is coded as "1 = winner" while Pakistan is coded "0 = loser"). In effect, this change allows us to control for the predictive power of the Epstein model in terms of determining if arms transfers affect the likelihood of winning—rather than the likelihood of winning impacting on the decision to participate (as was the case above).

Finally, the dependent variable is coded according to COW (1993), except where COW codes a war as a tie. In these cases (Iran-Iraq [1980],[69] Egypt-Israel [1969],[70] Israel-Syria [1982],[71] and China-Vietnam [1985]),[72] codings are done according to Arnold.[73]

Results: Testing Hypothesis 4-5, Do Arms Transfers Have an Effect on War Winners?

Tables 4.9 through 4.12 provide the results of the test of Model 3. The first of these tables shows that arms transfers are not a significant predictor of actual war outcomes in terms of who wins and loses. The prewar WEI capabilities ratio between the two states is, however, an effective indicator of which state will win the war.

**Table 4.9: The Effect of Arms Transfers on War Winners
with WEI Capabilities Ratio Controlled**

N	−2 Log Likelihood	Goodness of Fit	Model Chi-Square (significance)	
22	12.686	11.558	17.812 (0.0001)	
Variable	B	Std. Error	Significance	R
$\sum_{i=War} AT$ (WEI)	−3.3005	3.117	0.296	0.000
RATIO$_t$ (Aggregate WEI Values)	4.4717	2.217	0.043	0.260
Constant	−4.3326	2.076	0.037	—

Table 4.10 provides the breakdown between the predicted outcomes based on the model estimates and the actual outcomes. We see that even though the arms transfer variable is insignificant, the control variable ratio helps Model 3 to predict 81.82 percent of the cases correctly.[74] In effect, this model mistakenly predicts the outcome of one war (predicting that Uganda should overcome Tanzania), and in two others predicts wrongly that both sides should emerge either victorious (Iraq-Kuwait) or defeated (Iraq and Iran).

**Table 4.10: Classification Table for Arms Transfers Predicting
War Winners with WEI Capabilities Ratio Controlled**

Observed	Predicted		Percent Correct
	Loser	Winner	
Loser	9	2	81.82
Winner	2	9	81.82
		Overall	81.82

When we change the control variable from the WEI ratio to the POIs, we find very similar results. The statistics show that arms transfers are not significant predictors of war winners (even though one of our control variables are). However, because of the strength one of the iterations of the control variable POIs—winners—we must make some alterations in presenting the full analysis. A simple model with POI winners (the Epstein model's estimation of which side will win the war given the actual force values the first year of the war) predicting actual winners provides a success rate of over 90 percent (see note below for results).[75] The addition of the arms transfer variable not only indicates that the $\sum\limits_{i=War} AT$ is not statistically significant, but that its inclusion does not, however, degrade the ability of the model to predict actual winners (see Tables 4.11 and 4.12). If we switch to the other control variables, we find that neither the controls nor the $\sum\limits_{i=War} AT$ variable are statistically significant predictors of actual war winners.[76]

Table 4.11: The Effect of Arms Transfers on Actual War Winners with Predicted Winner Controlled

N	−2 Log Likelihood	Goodness of Fit	Model Chi-Square (significance)	
22	13.331	21.651	17.167 (0.0002)	
Variable	*B*	*Std. Error*	*Significance*	*R*
$\sum\limits_{i=War} AT$	−1.4005	6.199	0.820	0.000
Winner (POI of Epstein model)	4.5666	1.485	0.002	0.494
Constant	−2.2163	1.077	0.039	—

This model performs slightly differently than the WEI ratio controlled model in Table 4.10, as is seen in Table 4.12.

Table 4.12: Classification Table for Arms Transfers Predicting War Winners with Predicted Winner Controlled

Observed	Predicted Loser	Winner	Percent Correct
Loser	10	1	90.91
Winner	1	10	90.91
		Overall	90.91

An additional correctly predicted winner and loser is obtained with this model (compared to Table 4.10). Only one war is mistakenly predicted, that of Iraq-Kuwait, where the model wrongly predicts a Kuwaiti victory.[77]

This section provides additional insight into the phenomenon of arms transfer effects (or absence thereof) on war. Here, we sought to determine whether the amount of weapons imported by a state after the outbreak of war would have an effect on who wins the war. The theory indicated that arms transfers should make states with higher import values more likely to win a war, when the initial military balance between the states is controlled for. However, we find evidence in the analysis conducted here to indicate that this is not the case. It appears that arms transfers themselves *do not* predict actual war winners.

Models 4 and 5, Method

The final two models are examined by ordinary least squares regression. The questions concern war outcomes in terms of duration and casualties; i.e., whether arms transfers are associated with longer and bloodier wars. The models here assume that war is ongoing:

$$\frac{Milcap_a}{Milcap_b} + \sum_{i=War} AT \rightarrow Duration$$

(Model 4)

and

$$\frac{Milcap_a}{Milcap_b} + \sum_{i=War} AT \rightarrow Casualties$$

(Model 5)

In theoretical terms, there are competing expectations concerning the effect of arms transfers on war duration. The first is that arms transfers provide one state the necessary strength to close out the war, giving it enough of an edge over the opponent to ensure victory. Thus, one would expect to see that wars into which arms are transferred will end more quickly than those where they are not introduced. The alternative view is that arms transfers will replenish the supply of weapons of the weaker power, allow it to stay in the war, and possibly force a stalemate that will make the war longer.

In terms of casualties, however, there appears to be little argument as to what the effect of arms transfers is. Brzoska and Pearson indicate that arms transfers will lead to increased casualties as wars will "heat up," and conversely, the effect of embargoes (absence of arms transfers) will cause attrition rates to decline as the combatants become exhausted.[78]

Data

The data for Models 4 and 5 are much the same as above. The dependent variables, duration and casualties, will be coded from the "nation-months" and "battle deaths" variables of the COW (1993) data set. Unfortunately, because of the extreme duration (over 90 nation-months) and heavy losses incurred by both sides in the Iraq-Iran war (some 1.25 million), these cases must be eliminated because these figures are outliers some two entire orders of magnitude greater than any of the other wars in the data set.

Results: Testing Hypothesis 3-5, Do Arms Transfers Have an Effect on the Duration of War?

Our first task in determining whether arms transfers affect the duration of a war into which they are imported is to observe the correlation between the main variables (see Figure 4.2).

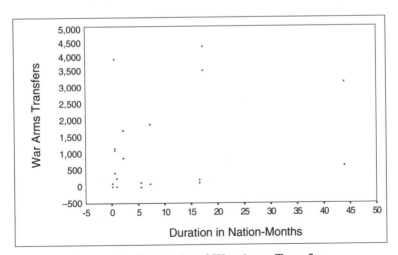

Figure 4.2: Scatterplot of War Arms Transfers and Duration in Nation-Months

The results from Pearson's correlations show that there is not a statistically significant relationship between war arms transfers and duration in nation-months ($R = 0.30$, $p = 0.19$)[79] When we proceed to an ordinary least squares[80] test of war arms transfers on duration, where we control for the ratio of military forces at the beginning of the war (measured in WEI as denoted above), we find that none of our predictor variables is statistically significant (war arms transfer *t-statistic* = 1.299, Signif. $t = 0.2112$).[81] Furthermore, our *F-statistic* indicates that we cannot reject the null hypothesis that none of the predictor variables is related at all to the duration data (*F-statistic* = 1.02818, Signif $F = 0.3789$). When we control for the POIs, we find very similar results to the ordinary least squares test (war arms transfers *t-statistic* = 1.081, Signif $t = 0.2981$; *F-statistic* = 0.31298, Signif $F = 0.8970$). However, it is noteworthy that in each of the tests, war arms transfers exhibits a positive, albeit statistically insignificant, relationship with duration in nation-months.[82]

This section examined the purported relationship between arms transfers during an ongoing war and the duration of the war. We found that arms transfers in general are not strongly related to, nor are a statistically significant predictor of, war duration. Thus, we may say that in border wars between contiguous states, at least, arms transfers do not seem to make wars last longer.

Results: Testing Hypothesis 4-5, Do Arms Transfers Have an Effect on Casualties in War?

In order to test Hypothesis 4-5 concerning casualties, we may again want to observe the simple correlation between $\sum_{i=War} AT$ and casualties (see Figure 4.3).

The two-tailed Pearson's correlation coefficient between $\sum_{i=War} AT$ and battle deaths is .0532 and is not statistically significant (nor is the correlation between battle deaths and any of our WEI or POI controls significant). When we control for the military capabilities ratio (prewar WEI scores) in a multiple ordinary least squares regression[83] model of arms transfers on battle deaths, we find again no relationship (*t-statistic* = 0.237, Signif. $t = 0.8151$; *F-statistic* = 0.34, Significance = 0.71).[84] In a similar model that controls for the several POI estimations, we again find the same absence of a relationship (*t-statistic* = 0.500, Signif. *t* = 0.6246; *F-statistic* = 0.68, Significance = 0.64).

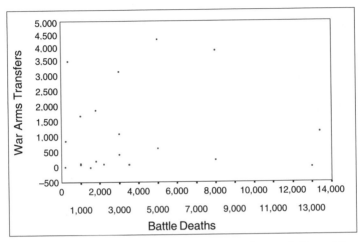

Figure 4.3: Scatterplot of War Arms Transfers and Casualties

Based on the consistency of these results, we must reject the hypothesis that arms transfers are statistically related to, or have a statistical effect on, casualties.

Conclusion: The Effects of Arms Transfers on War Involvement and Outcomes

In this chapter, we set out to explore the relationship between arms transfers and war involvement and war outcomes. This was done in discreet steps; first by assessing the literature on international conflict and arms transfers on the subject, then by deriving hypotheses from the literature for tests that were conducted. The tests themselves took three steps: determining whether the military balance was indeed an intervening variable in the model of whether arms transfers affected war involvement; assessing the extent of the direct relationship between arms transfers and war involvement; and finally empirically examining if arms transfers were related to war outcomes.

The examination of Hypothesis 4-1 (*Arms transfers make a significant impact on the military balance between rival states*) used military modeling techniques to make what Bennett and Stam call "ex ante predictions," predictions based upon information that a leader would have *prior to* engaging in war.[85] However, whereas their model was able to make somewhat accurate predictions (mean error of about 1.5 *years*) as to the duration (only) of interstate wars, the methods of measurement used here to make ex ante predictions (derived from the dynamic, aggregate weighted effectiveness index [WEI] of Raymond and the dynamic force attrition models of Epstein) gave military capabilities by a method actually used by military planners and strategists and in outcome indicators in terms of casualties, front displacement, winners, and duration derived from military models of war. Based on the analysis of the data generated for 22 test cases of war between contiguous states and 20 control cases of matched contiguous states that did not engage in war, we were encouraged to reject Hypothesis 4-1 (see Table 4.13). The ex ante measures used by military modelers did not generate results that indicated that there was a statistical difference made by arms transfers on the military balance.

Due to the findings concerning Hypothesis 4-1, we declined to discuss in this chapter whether arms transfers affect war involvement by changing the military balance between states (denoted Hypotheses 4-2 and 4-3 on page 95), and instead proceeded immediately to the examination of whether arms transfers directly affect the probability of a state being involved in a war with a neighbor. The findings for this hypothesis, again summarized in Table 4.13, indicate that there is a statistically significant, positive relationship between arms transfers and war involvement. The models investigated there were able to correctly predict about 80 percent of war involvement in the test and control sets of our data. Based on the strength and consistency of this evidence, we are encouraged to accept Hypothesis 4-4.

Finally, we sought to determine whether arms transfers played a role in determining war outcomes. The latter were measured in terms of the following indicators: who wins the war, the length of the war, and the number of casualties of the war. Our examination did not reveal that arms transfers themselves were suitable predictors of war outcomes in a statistical sense (although our use of military science techniques to predict war winners was successful in making correct predictions over 90 percent of the time). Thus, we were forced to reject Hypothesis 4-5, that arms transfers predicted war outcomes.

**Table 4.13: Summary of Empirical Tests Performed in Chapter 4,
Hypotheses, Findings, and Recommendations**

Hypothesis	Findings	Recommendation
4-1: Arms transfers make a significant impact on the military balance between rival states.	Difference between military balances with and without arms transfers in 22 test and 20 control cases statistically indistinguishable.	Reject Hypothesis 4-1.
4-4: Given a contiguous dyad and an established ratio of military capability between the states of the dyad, arms transfers increase the probability of war involvement between the dyad members.	Arms transfers in the three-year period prior to war outbreak (or hypothetical outbreak, for control cases) are positive and statistically significant predictors of war involvement. Models predict correctly over 80 percent of cases.	Accept Hypothesis 4-4.
4-5: Weapons transfers into a contiguous borders conflict dyad are associated with measurable, direct, and significant alterations of the outcomes (winners, duration, and casualties) of the war.	Arms transfers that take place during a war do not predict war outcomes (winners, duration, or casualties) to commonly accepted statistical significance.	Reject Hypothesis 4-5.

The implications of these findings are manifest. Arms transfers quite possibly play an important, direct role in leaders' decisions regarding war involvement. According to the SIPRI theory, this is because the amount of resources devoted to the military, as evidenced by levels of weapons imports, makes decision makers more aware of military prowess when it comes to influencing foreign policy. In turn, the militarization of foreign policy makes these countries much more likely to become involved in wars with their neighbors. Subsequently, however, we lose the theoretical relationship most often posited between arms transfers and war outcomes. In the case of winners, arms transfers did not seem to matter, as they were in effect unable to overcome the predictive power of overall military capabilities on the first day of the war. This concurs with our previous finding that arms transfers do not, in actuality, alter the military balance between states. In effect, this study gives credence to the oft-found relationship that when war breaks out, God is on the side of the big (or in this case, more capable) battalions when it comes to war outcomes.

Furthermore, the theoretical ambiguity surrounding the effects of war arms transfers on the duration of war are again encountered. This study showed no relationship. In effect, this is the same as saying that all of the theoretical propositions concerning this relationship are neither proved nor disproved. In specific cases, we may be able to predict the effect of arms transfers on war duration, as claimed by Pearson, Brzoska, and Crantz and by Brzoska and Pearson, but there seems to be no generally detectable effect across the population of cases. We simply do not know whether there is truly "no relationship" between arms transfers and duration, or whether there are relationships that are, in effect, canceling each other out; some arms transfers do allow a side to close out the war while others replenish the strength of the belligerents and allow them to struggle on. Indeed, for the policy maker there appear to be no one-armed prognosticators on the effects of arms transfers on war duration any more than there are one-armed economists to explicate economic policy.

Finally, war casualties seem to be unrelated to the amount of weapons transferred into a conflict. This is against the prevailing theory offered most cogently by Brzoska and Pearson, who found that arms transfers causing war to be bloodier was a "general finding." The alternative logic to this has in the past been neglected, namely that arms transfers are not related to the casualties of a war perhaps because it is the *placement* of their destructive power, not its *existence*, that cause wars to be bloody. In other words, the potential destructive power of any weapon is a function of whether it goes off in a desolate, uninhabited location or a highly developed area with dense population. In the one case, the weapon will not be related to bloodiness. In the other, it stands a much better chance of being so. This logic applies to the battlefield as well, as will be discussed further below. In addition, even if having more weapons allowed a state to increase the tempo of operations in a war, doing so would not necessarily cause more casualties. The enemy could trade space for time or to conserve forces (breaking contact serving to decrease likelihood of casualties).

Alternatively, imported weapons may have an asymmetric effect on the types of capabilities that are survivable on the battlefield, serving to decrease the casualties in a war if, for instance, the imported weapon is a dedicated fighter aircraft which by virtue of its ability to deny airspace to enemy bombers *decreases* the casualty rate of its owner.[86] Finally, there is the possibility that casualty rates in wars are elastic in the sense that under conditions of greater concentration of firepower, forces will disperse in order to decrease the likelihood of incurring casualties, whereas under conditions of less firepower, forces will concentrate and thus become more likely to suffer massive losses.[87] The evidence generated by the empirical tests in this chapter indicates that more attention needs to be devoted to these alternative theories of the effects of weapons on war casualties.

To conclude, this chapter provides interesting observations about the relationship between weapons and war. The importation of weapons allows us to make some modest predictions concerning the likelihood of war involvement of the recipient state. The findings tell us also that the statement by the U.S. executive branch to Congress (used as a header to Chapter 3) in the notification of pending weapons sales abroad may be nothing more than self-serving rhetoric designed to allow American companies to export weapons abroad, which in turn allows for longer production runs and lesser per-unit costs for the U.S. government in buying the same weapons. That the sale of weapons does not change the military balance in the region may be trivial if Creveld is correct and leaders do not uniformly make "rational" calculations in their decisions to go to war. If so, then at worst such statements by the executive branch are "only" misleading to Congress and therefore the American people. This chapter does not provide a fully specified alternative model upon which to make decisions regarding the sale of weapons abroad (although this is a potentially fruitful extension), but it does indicate that decision makers have some evidence to re-examine such policies, and also that they have the tools with which their determinants can be more clearly analyzed when considering weapons sales.

Notes

1. Lincoln Bloomfield and Amelia Leiss, *Controlling Small Wars; A Strategy for the 1970's* (New York: Knopf, 1969).

2. Martin van Creveld, *On Future War* (London: Brassey's, 1991).

3. See Ronald Sherwin and Edward Laurance, "Arms Transfers and Military Capability: Measuring and Evaluating Conventional Arms Transfers," *International Studies Quarterly* 23 (1979): 360–89; Ronald Sherwin, "Controlling Instability and Conflict through Arms Transfers: Testing a Policy Assumption," *International Interactions* 10 (1983): 65–99; Donald Sylvan, "Consequences of Sharp Military Assistance Increases for International Conflict and Cooperation," *Journal of Conflict Resolution* 20 (1976): 609–36; William Baugh and Michael Squires, "Arms Transfers and the Onset of War Part I: Scalogram Analysis of Transfer Patterns," *International Interactions* 10 (1983): 39–63; William Baugh and Michael Squires, "Arms Transfers and the Onset of War Part II: Wars in Third World States,

1950–65," *International Interactions* 10 (1983): 129–41; and Stephanie Neuman, "The Role of Military Assistance in Recent Wars," in *The Lessons of Recent Wars in the Third World, Volume II*, ed. Stephanie Neuman and Robert Harkavy (Lexington, MA: D.C. Heath, 1986), 115–56.

4. The term *arms races* often used in this literature is misleading as the authors rarely consider acquisition or number of weapons in their calculations. Instead, they typically use proxies such as rate of change in military spending. Michael Wallace, *War and Rank Among Nations* (Lexington, MA: D.C. Heath, 1973); Michael Wallace, "Arms Races and Escalation," *International Studies Quarterly* 26 (1982): 37–56; and Michael Wallace, "Arms Races and Escalation: Some New Evidence," *Journal of Conflict Resolution* 23 (1982): 3–16.

5. Paul Diehl, "Arms Races and Escalation: A Closer Look," *Journal of Peace Research* 20 (1983): 205–12; Paul Diehl, "Arms Races to War: Testing Some Empirical Linkages," *Sociological Quarterly* 96 (1985): 331–49; Paul Diehl and Jean Kingston, "Messenger or Message?: Military Buildups and the Initiation of Conflict," *Journal of Politics* 49 (1987): 801–13.

6. See, for example, Eric Weede, "Arms Races and Escalation: Some Persisting Doubts," *Journal of Conflict Resolution* 24 (1980): 285–88; and Robert Powell, "Stability and the Distribution of Power," *World Politics* 48 (1996): 239–67.

7. Hans Morgenthau, *Politics among Nations: The Struggle for Power and Peace* (New York: Alfred Knopf, 1948).

8. See George Modelski and William Thompson, *Seapower in Global Politics, 1494–1993* (Seattle: University of Washington Press, 1988); A.F.K. Organski and Jacek Kugler, *The War Ledger* (Chicago: University of Chicago Press, 1980); Robert Gilpin, "The Theory of Hegemonic War," in *The Origin and Prevention of Major Wars*, ed. R. Rotberg and Theodore Rabb (Cambridge: Cambridge University Press, 1989).

9. Bruce Bueno de Mesquita, *The War Trap* (New Haven, CT: Yale University Press, 1981); and Bruce Bueno de Mesquita and David Lalman, *War and Reason* (New Haven, CT: Yale University Press, 1992).

10. Bueno de Mesquita bases this calculation on the "relative power" of the adversaries, measured in terms of "military, industrial and demographic" capability. The Correlates of War indices of state power that he and others use include such items as military spending, military manpower, population, urban population, and iron and steel production. For an explanation of these indicators, see Melvin Small and J. David Singer, Resort to Arms: International and Civil Wars, 1816-1980 (Beverly Hills: Sage, 1982).

11. T.V. Paul, *Asymmetric Conflicts: War Initiation by Weaker Powers* (New York: Cambridge University Press, 1994).

12. As alluded to in the text, Paul focuses his study of "asymmetric" wars where the weak attack the strong. It is a study, therefore, of a special subset of wars. His main analytical focus is on the strategic concept of *fait accompli*, or the offensive-defensive/deterrent. Paul, *Asymmetric Conflicts*, 1994.

13. In their respective formulations, Rosen and Organski and Kugler find that the side best able to mobilize its domestic resources wins. Organski and Kugler incorporate foreign aid in their formulation of power, but do not single out the most important aspect of this aid—weapons. These studies both attempt to incorporate intangible aspects of military power (such as leadership, willingness to suffer, ability to extract resources, etc.) when exploring conflict outcomes. Steven Rosen, "The Proliferation of New Land-Based Technologies: Implications for Local Military Balances," in *Arms Transfers in the Modern World*, ed. Stephanie Neuman and Robert Harkavy (New York: Praeger, 1979), 109–30; Organski and Kugler, *The War Ledger*, 1980.

14. Allan Stam, *Win, Lose, or Draw: Domestic Politics and the Crucible of War* (Ann Arbor: University of Michigan Press, 1996). See also D. Scott Bennett and Allan Stam, "The Duration of Interstate Wars, 1816–1985," *American Political Science Review* 90 (1996): 239–57.

15. The COW indices of state power include such items as military spending, military manpower, population, urban population, and iron and steel production. For an explanation of these indicators, see Small and Singer, *Resort to Arms,* 1982.

16. See, for example, Organski and Kugler, *The War Ledger*, 1980; Henk Houweling and Jan Siccama, *Studies of War* (Dordrecht, The Netherlands: Martinus Nijhoff, 1988); Richard Merritt and Dina Zinnes, "Validity of Power Indices," *International Interactions* 14 (1988): 141–51; William Moul, "Measuring the 'Balances of Power': A Look at Some Numbers," *Review of International Studies* 15 (1989): 101–21; John O'Neal, "Measuring the Material Base of the Contemporary East-West Balance of Power," *International Interactions* 15 (1989): 177–96; and Richard Stoll and Michael Ward, *Power in World Politics* (Boulder, CO: Lynne Rienner, 1989).

17. Hans Rattinger, "From War to War to War: Arms Races in the Middle East," *International Studies Quarterly* 20 (1976): 501–31; Ronald Sherwin and Edward Laurance, "Arms Transfers and Military Capability: Measuring and Evaluating Conventional Arms Transfers," *International Studies Quarterly* 23 (1979): 360–89.

18. Paul Huth, D. Scott Bennett, and Christopher Gelpi, "System Uncertainty, Risk Propensity, and International Conflict among the Great Powers." *Journal of Conflict Resolution* 36 (1992): 478–517.

19. An interesting and rigorous element of this literature is the attempt by Sanjian to develop formal, probabilistic models of the arms transfer decision-making process (from the supplier's side). Gregory Sanjian, *Arms Transfers to the Third World: Probability Models of Superpower Decisionmaking* (Boulder, CO: Lynne Rienner, 1987); Gregory Sanjian, "Arms Export Decision-Making: A Fuzzy Control Model," *International Interaction* 14 (1988): 243–65; Gregory Sanjian, "Fuzzy Set Theory and U.S. Arms Transfers: Modeling the Decision-Making Process," *American Journal of Political Science* 32 (1988): 1018–46; Gregory Sanjian, "Great Power Arms Transfers: Modeling the Decision-Making Processes of Hegemonic, Industrial, and Restrictive Exporters," *International Studies Quarterly* 35 (1991): 173–93. See also, Stephen Kaplan, "U.S. Arms Transfers to Latin America, 1945–1974," *International Studies Quarterly* 19 (1975): 399–431, who looks at rational strategic, bureaucratic politics, and executive persuasion models in the U.S. case.

20. Christian Catrina, "Main Directions of Research in the Arms Trade," *Annals of the American Academy of Political and Social Science* 535 (1994): 190–205.

21. See, for example, Stephanie Neuman, *Military Assistance in Recent Wars* (New York: Praeger, 1986); Robert Harkavy, "Recent Wars in the Arc of Crisis: Lessons for Defense Planners," in *Defense Planning in Less-Industrialized States: The Middle East and South Asia,* ed. Stephanie Neuman (Lexington, MA: D.C. Heath, 1984), 275–300; William Staudenmaier, "Iran-Iraq (1980–)," in *The Lessons of Recent Wars in the Third World, Volume I,* ed. Robert Harkavy and Stephanie Neuman (Lexington, MA: D.C. Heath, 1985), 211–38; and John Lambelet, "A Numerical Model of the Anglo-German Dreadnought Race," *Peace Science Society (International), Papers* 24 (1975): 29–48.

22. See also Rattinger, "From War to War to War," 1976; and Baugh and Squires, "Arms Transfers and the Onset of War Part I," 1983.

23. See also Baugh and Squires, "Arms Transfers and the Onset of War Part I," 1983; Joshua Epstein, *The Calculus of Conventional War: Dynamic Analysis without Lanchester Theory* (Washington, D.C.: Brookings Institution, 1985); Rattinger, "From War to War to War," 1976; William Mako, *US Ground Forces and the Defense of Central Europe* (Washington, D.C.: Brookings Institution, 1983); and A. D. Raymond, "Assessing Combat Power: A Methodology for Tactical Battle Staffs," (Fort Leavenworth, KS: Army Command and General Staff College, 1991, for similar efforts).

24. David Kinsella, "Conflict in Context: Arms Transfers and Third World Rivalries during the Cold War," *American Journal of Political Science* 38 (1994): 557–81; and David Kinsella and Herbert Tillema, "Arms and Aggression in the Middle East," *Journal of Conflict Resolution* 39 (1995): 306–29.

25. Philip Schrodt, "Arms Transfers and International Behavior in the Arabian Sea Area," *International Interactions* 10 (1983): 101–27. Frederic Pearson, Michael Brzoska, and Christopher Crantz, "The Effect of Arms Transfers on Wars and Peace Negotiations," in SIPRI, *SIPRI Yearbook 1992, Armaments and Disarmament* (Oxford: Oxford University Press, 1992); and Michael Brzoska and Frederic Pearson, *Arms and Warfare: Escalation, De-escalation, and Negotiations* (Columbia: University of South Carolina Press, 1994).

26. Pearson, Brzoska, and Crantz, "The Effect of Arms Transfers on Wars and Peace Negotiations," 1992; Brzoska and Pearson, *Arms and Warfare,* 1994.

27. Pearson, Brzoska, and Crantz, "The Effect of Arms Transfers on Wars and Peace Negotiations," 1992, 399; and similarly, Brzoska and Pearson, *Arms and Warfare,* 1994, 214.

28. Pearson, Brzoska, and Crantz, "The Effect of Arms Transfers on Wars and Peace Negotiations," 1992, 399; and similarly, Brzoska and Pearson, *Arms and Warfare,* 1994, 216.

29. Brzoska and Pearson, *Arms and Warfare,* 1994, 216.

30. SIPRI, *The Arms Trade with the Third World* (Stockholm: Almquist and Wiksell, 1971).

31. Catrina, "Main Directions of Research in the Arms Trade," 1994.

32. Decline of military power over distance is explored in Kenneth Boulding, *Conflict and Defense: A General Theory* (New York: Harper and Row, 1963); and developed further in Bueno de Mesquita, *The War Trap*, 1981.

33. For a more complete discussion of logistic regression, see Chapter 3.

34. Because of gaps in our data concerning military balances, analysis of border wars that occurred prior to 1969 is not practical. For the same reason, the Football War (1969) between El Salvador and Honduras cannot be analyzed. Therefore, these border wars have been eliminated from the data set used here, COW (1993—interstate wars only).

35. In accordance with Kenneth Meier and Jeffrey Brudney, *Applied Statistics for Public Administration*, rev. ed. (Pacific Grove, CA: Brooks/Cole Publishing Company, 1987), "the objective of matching is to create experimental and control groups that are as similar as possible on selected characteristics."

36. See Raymond, "Assessing Combat Power," 1991; Trevor Dupuy, *Numbers, Predictions and War: Using History to Evaluate Combat Factors and Predict the Outcome of Battles* (Indianapolis: Bobbs-Merrill, 1979); Trevor Dupuy, *Analysis of Factors That Have Influenced Outcomes of Battles and Wars: A Data Base of Battles and Engagements, Final Report,* 6 vols. (Dunn Loring, VA: Historical Evaluation and Research Organization, 1983); and Leslie Callahan, Jr., "The Need for a Multidisciplinary Modeling Language in Military Science and Engineering," in *Modeling and Simulation of Land Combat*, ed. Leslie Callahan, Jr. (Atlanta: Georgia Tech Research Institute, 1983).

37. See Mako, *US Ground Forces and the Defense of Central Europe*, 1983; U.S. Congressional Budget Office (hereafter CBO), *U.S. Ground Forces and the Conventional Balance in Europe* (Washington, D.C.: U.S. Government Printing Office, 1988).

38. The only attempt that I have discovered in the arms transfers literature to either predict the outbreak of wars or to combine both quality and quantity of weapons sold is Baugh and Squires, "Arms Transfers and the Onset of War Part I," 1983; and Baugh and Squires, "Arms Transfers and the Onset of War Part II," 1983. Unfortunately, these efforts and methodological improvisations were not combined into a coherent investigation of the effect of arms transfers on war. For other seminal attempts to derive qualitative and quantitative values for weapons, see Dupuy, *Numbers, Predictions and War,* 1979; Dupuy, *Analysis of Factors That Have Influenced Outcomes of Battles and Wars,* 1983; and James Dunnigan, *How to Make War* (New York: William Morrow, 1988).

39. See Raymond, "Assessing Combat Power," 1991; CBO, *U.S. Ground Forces and the Conventional Balance in Europe*, 1988; and Mako, *US Ground Forces and the Defense of Central Europe*, 1983.

40. Dupuy, *Numbers, Predictions and War,* 1979; Raymond, "Assessing Combat Power," 1991, suggests a similar ratio in his study.

41. A quick assessment of payloads and accuracy assures us that this is true, albeit conservative, because this relationship indeed conforms to the condition that for destructive force, an exponential function of payload is necessary to equal a unit improvement in accuracy. In an illustrative case, a MIG-23 fighter-bomber has an attack value of 24, which is well in excess of the comparable value of 4 for a MI-24 attack helicopter, where these attack values, according to Dunnigan, *How to Make War*, 1988, are normalized for accuracy.

42. See Raymond, "Assessing Combat Power," 1991.

43. In the estimation of aggregate WEIs, a state's military personnel in the regular Army are coded in the following manner. A state's total regular army forces are divided by 30 (in order to achieve a consistent unit of scale with Raymond), and then coded according to Raymond's "Type A infantry;" i.e. 1.3 Weapon Value. For nonregular forces (reserves and militias), we perform the same division and then code them according to Raymond's "Type C infantry;" i.e., 0.8 Weapon Value.

44. As noted above, military spending per soldier has real weaknesses as a measure of military capability when considered in isolation from the quantity and quality of a military's weapons on hand. The formulation of military capability used here, of course, accounts for both of these separately. An alternative measure would be to review the historical literature concerning the military establishment of a given country (especially its performance during wars) and provide a rating based upon this review and assessment (for an interesting means of evaluation along these lines, see Dupuy, *Numbers, Predictions and War*, 1979). The weakness of this alternative is that making a *post-hoc* assessment of a state's military prowess based on its performance in the war under consideration is a questionable means of determining how a decision maker may assess the military capability *prior to* the war. For a discussion of this last point, see Bueno de Mesquita, *The War Trap*, 1981.

45. The most comprehensive account of the development and usage of Lanchester equations (and variants) is James Taylor, *Lanchester Models of Warfare*, 2 vols. (Arlington, VA: Operations Research Society of America, 1983). See also James Taylor, *Force-on-Force Attrition Modeling* (Arlington, VA: Operations Research Society of America, 1981); Jerome Bracken, Moshe Kress, and Richard Rosenthal, eds., *Warfare Modeling* (Danvers, MA: John Wiley and Sons, Inc., 1995); and Charles Anderton, "Toward a Mathematical Theory of the Offensive/Defensive Balance," *International Studies Quarterly* 36 (1992): 75–100.

46. According to Bruce Fowler, "A *Perestroika* Program for Modeling and Simulation Tools and Techniques," in *Proceedings of the 1991 Callaway Workshop: Modeling, Simulation, and Gaming for Restructuring the U.S. Armed Forces,* ed. Leslie Callahan, Jr., and Ross Gagliano (Pine Mountain, GA: U.S. Army Missile Command, 1992), warfare has long been dominated by the "attrition mechanics" of a type investigated by Lanchester in 1916 with *Aircraft in Warfare: The Dawn of the Fourth Arm*, and concurrently in Russia with Mikhail Osipov in 1915, with "The Influence of the Numerical Strength of Engaged Sides on their Casualties," published in *Voenniy Sbornik* (*Military Collection*) and translated by Helmbold and Rehm in Bracken, Kress, and Rosenthal, eds., *Warfare Modeling*, 1995, 289–344. Since Lanchester is much better known in the West and translations of Osipov have

yet to be fully analyzed and digested by Western researchers, the discussion here is limited to the Englishman. According to Lanchester's logic, an assessment of a state's military capability versus an opponent's takes the following form: $dR/dt = -bB$ (the instantaneous rate of change of the strength [technology as a function of size and effectiveness] of Red's forces equals a constant multiplied by Blue's strength), while $dB/dt = -rR$ is the same for Blue's forces, so that $dR/dB = bB/rR$. This implies that the casualty exchange ratio between the two forces would exhibit a linear relationship to the two sides' force ratio during a potential conflict (for other treatments, see Taylor, *Lanchester Models of Warfare*, 1983; Taylor, *Force-on-Force Attrition Modeling*, 1981; John Lepingwell, "The Laws of Combat?: Lanchester Reexamined," *International Security* 12 (1987): 89–139; and Anderton, "Toward a Mathematical Theory of the Offensive/Defensive Balance," 1992.

47. For reviews of these results, see U.S. General Accounting Office ((hereafter GAO), *Models, Data, and War: A Critique of the Foundation for Defense Analysis*, PAD-80-21 (Washington, D.C.: U.S. Government Printing Office, 1980); Robert Helmbold, "Some Observations on the Use of Lanchester's Theory for Prediction," *Operations Research* 12 (1964): 778–81; Herbert Weiss, "Combat Models and Historical Data: The U.S. Civil War," *Operations Research* 14 (1966): 228–48; J.H. Engle, "A Verification of Lanchester's Law," *Operations Research* 2 (1954): 12–48; and D. Willard, *Lanchester as Force in History: An Analysis of Land Battles of the Years 1618–1905*, RAC-TP-74 (Bethesda, MD: Research Analysis Corporation, 1962).

48. See Epstein, *The Calculus of Conventional War*, 1985; William Kaufmann, "The Arithmetic of Force Planning," in *Alliance Security: NATO and the No-First-Use Question*, ed. John Steinbruner and Leon Sigal (Washington, D.C.: Brookings Institution, 1983); and Raymond, "Assessing Combat Power," 1991. For examples of the use of force planning models in other countries, see Yu Chuyev *Fundamentals of Operations Research in Combat Materiel and Weaponry*, 2 vols. (Wright Patterson AFB, OH: Foreign Technology Division, 1968); and Yu Chuyev and Yu Mikhaylov, *Forecasting in Military Affairs: A Soviet View* (Washington, D.C.: U.S. Government Printing Office, 1980).

49. For examples of this literature, see Paul Davis, "Variable-Resolution Combat Modeling," in *Proceedings of the 1991 Callaway Workshop*, ed. Callahan and Gagliano, 1992; Dean Hartley, "Mathematical Modeling of Historical Combat Data," in *Proceedings of the 1991 Callaway Workshop*, ed. Callahan and Gagliano, 1992; Kaufmann, "The Arithmetic of Force Planning," 1983. Bracken, Kress, and Rosenthal, *Warfare Modeling*, 1995; and R. Shephard, D. Hartley, P. Haysman, and L. Thorpe, *Applied Operations Research: Examples from Defense Assessment* (New York: Plenum Press, 1988).

50. A somewhat lengthy discussion of the merits of force-attrition models and their role in the field of strategic studies includes Joshua Epstein, "The 3:1 Rule, the Adaptive Dynamic Model, and the Future of Security Studies," *International Security* 13 (1989): 90–127; John Mearsheimer, "Assessing the Conventional Balance: The 3:1 Rule and Its Critics," *International Security* 13 (1989): 54–89; Lepingwell, "The Laws of Combat?" 1987; and Barry Posen, "Measuring the European Conventional

Balance: Coping with Complexity in Threat Assessment," *International Security* 9 (1984–5): 56–81. Other parts of this discussion are Joshua Epstein, Kim Holmes, John Mearsheimer, and Barry Posen, "The European Conventional Balance," *International Security* 12 (1988): 152–202; and John Mearsheimer, Barry Posen, Eliot Cohen, Steven Zaloga, Malcolm Chalmers, and Lutz Unterseher, "Correspondence," *International Security* 13 (1989): 128–79.

51. Epstein, *The Calculus of Conventional War,* 1985. The CBO illustrates the importance of such assessments in such statements as, for example, "Deterrence of war in Europe—or if necessary, its successful prosecution—depends in part on the balance of conventional forces" between NATO and the Warsaw Pact. CBO, *U.S. Ground Forces and the Conventional Balance in Europe,* 1988.

52. Geoffrey Blainey, *The Causes of War,* 3d ed. (New York: Free Press, 1988), 122. See also Creveld, *On Future War,* 1991.

53. George Quester, "Six Causes of War," in George Quester, *The Future of Nuclear Deterrence* (Lexington, MA: Lexington Books, 1987).

54. See Joshua Epstein, *Measuring Military Power: The Soviet Air Threat to Europe* (Princeton, NJ: Princeton University Press, 1984); Epstein, *The Calculus of Conventional War,* 1985; Joshua Epstein, *Strategy and Force Planning: The Case of the Persian Gulf* (Washington, D.C.: Brookings Institution, 1987); and Joshua Epstein, *Conventional Force Reductions: A Dynamic Assessment* (Washington, D.C.: Brookings Institution, 1990). Epstein's model is modified for use in analyzing the conventional balance between NATO and the Warsaw Pact in CBO, *U.S. Ground Forces and the Conventional Balance in Europe,* 1988.

55. Epstein, *The Calculus of Conventional War,* 1985, 17–18.

56. We can also predict who would win a conflict by using Epstein's models if we allow one side's force strength to go to some predefined floor level and define that floor strength as a "loss." For the analysis here, this is defined very conservatively at 10 percent of the beginning strength of the attacker, following Epstein, *The Calculus of Conventional War,* 1985, as long as there is concurrently a condition of no front movement, or zero strength for the defender. While this is ahistorical (it is hard to find evidence of any military unit maintaining an attack after having taken more than 50 percent casualties), it must suffice until we have a truly dynamic model that would allow counterattacks by the initial defender and thus a more historical outcome.

57. Epstein, *The Calculus of Conventional War,* 1985; and Epstein, *Conventional Force Reductions,* 1990. The alteration from Epstein is required because his model was designed for the study of war between major Warsaw Pact and NATO units. Because the basic unit of analysis in Epstein's model, the armored division equivalent (or ADE) is virtually meaningless for states such as Vietnam or Uganda, and it is not useful for this study. The specifics of the alterations are noted in Appendix 4. It is important to note that these alterations are consistent with the WEI/POI system described here.

58. Raymond, "Assessing Combat Power," 1991.

59. See CBO, *U.S. Ground Forces and the Conventional Balance in Europe*, 1988; and Mako, *US Ground Forces and the Defense of Central Europe*, 1983.

60. For the methodological problems in loss of model accuracy caused by the aggregation of data, see Davis, "Variable-Resolution Combat Modeling," 1992; and Richard Hillestad and Mario Juncosa, "Cutting Some Trees to See the Forest: On Aggregation and Disaggregation in Combat Models," in *Warfare Modeling*, ed. Bracken, Kress, and Rosenthal, 1995. The point here is that with better data, the model used for this analysis should improve.

61. The aggregation of the data for this study suffers from a lack of precision that a decision maker would (arguably) have before making estimations of war outcomes either in an ex ante fashion, or with attrition models. Besides the probable inconsistencies in *The Military Balance* data set, there are other hindrances to utilizing the full potential of the military model used as well. The types of information that would improve predictive power are: specifics on types of terrain so that variation may be included; existence of defensive fortifications; orders of battle for both sides; mobilization and reinforcement schedules; and operational strategy (for examples of modeling with these types of information, see CBO, *U.S. Ground Forces and the Conventional Balance in Europe*, 1988; Mako, *US Ground Forces and the Defense of Central Europe*, 1983; and Epstein, *Conventional Force Reductions*, 1990. If this information is available, it can be easily introduced into a military science model. Because in this study we suffer from lack of these types of information, certain problems of aggregation are unavoidable; i.e., the necessity of simulating a randomly selected strategic parameter such as an "all-out" attack by one country using all of the forces available [when number of borders is controlled for]).

62. Further analysis of the strengths and weaknesses of SIPRI and ACDA data can be found in Michael Brzoska, "Arms Transfer Data Sources," *Journal of Conflict Resolution* 26 (1982): 77–108; Edward Fei, "Understanding Arms Transfers and Military Expenditures: Data Problems," in *Arms Transfers in the Modern World*, ed. Stephanie Neuman and Robert Harkavy (New York: Praeger, 1979), 37–48; and Edward Kolodziej, "Measuring French Arms Transfers," *Journal of Conflict Resolution* 23 (1979): 195–227. The best work on the UN register of conventional weapons transfers is Edward Laurance, *Arms Watch: SIPRI Report on the First Year of the UN Register of Conventional Arms* (Oxford: Oxford University Press, 1993).

63. SIPRI also provides data on radar and guidance systems, as well as naval weapons. Neither of these two categories are included in the analysis here.

64. Complete results are available in Appendix 4.

65. As implied in the text, these findings do not indicate that there is no absolute effect of arms transfers on the military balance. Weakness in the data set surrounds the estimation technique used to derive the military force ratios. These data are imperfect in part because of the assumptions inherent in the formula used to derive the amount of the state's total force that can be applied to any given border-sharing adversary; i.e., that the country would need to leave equal portions of its total force on its other borders. This is an ahistorical assumption for individual cases, but because we are attempting a general model it must be acceptable (or some variation

thereof). A historical investigation of what is currently estimated would improve the veracity of the model, but is at this time unavailable in the unclassified literature.

66. Because the duration POI did reach a level of significance in the difference of means test, examinations of this and the other indicators, along the lines of Hypotheses 4-2 and 4-3, are available in Appendix 4.

67. These terrain types are listed in Appendix 4.

68. This technique was described and used in Chapter 3.

69. Iran is coded as the winner of the Iraq-Iran conflict for the following reasons: (1) Iraq attacked, and thus challenged the *status quo antebellum*; (2) with the United Nations' sponsored cease-fire that ended the hostilities in 1988, Guy Arnold (*Wars in the Third World Since 1945* [London: Cassell, 1995], 257) observes that the *status quo antebellum* was restored. Thus, Iran's successful defense of the *status quo antebellum* constitutes a victory.

70. Arnold, *Wars in the Third World Since 1945*, 1995, 325, denotes that the cease-fire brokered by the U.S. in 1970, which ended the desultory fighting in the Sinai, recognized an important victory by Israel because it allowed Israel to trade land (which it probably had no interest in holding—the Sinai) for *de facto* recognition of the state of Israel by the Arabs. As we know that the UN Resolution 242 demilitarized the Sinai (providing peacekeepers and concrete verification measures), the land "lost" by Israel must receive considerably less weighting than the political "victory" gained by recognition by two of its Arab enemies. Based on this reasoning, the 1969 "war" is coded as an Israeli victory.

71. Arnold, *Wars in the Third World Since 1945*, 1995, 338, notes that Israel obtained all of its objectives in the invasion of southern Lebanon in 1982 *and* inflicted a military defeat on Syria at the same time. Thus, Israel won.

72. In the Chinese-Vietnamese border wars during the mid-1980s (represented by COW as a 1985–87 war that ended in a tie), Arnold, *Wars in the Third World Since 1945*, 1995, 274, observes that the Chinese eventually obtained their primary objective—the Vietnamese withdrawal from Kampuchea. Since this indeed occurred in 1989, it seems that China obtained its objective. Thus, it is coded as having won its border dispute with Vietnam.

73. Arnold, *Wars in the Third World Since 1945*, 1995.

74. A naive model would, of course, predict correctly 50 percent of the time.

75. These are the results of the POI Winners predicting Actual Winners model:

Variable	B	Std. Error	Significance	R
Winner (Prediction of Epstein model)	4.6051	1.483	0.001	0.500
Constant	−2.3026	1.048	0.028	—

The classifications of the resultant predictions indicate that the model is successful 90.91 percent of the time. In effect, it mistakenly predicts one winner that really lost, and one loser that won in actuality. Model chi-square is significant at the .0000 level. Log likelihood is 13.404, and goodness of fit is 22.0.

76. The model of arms transfers predicting actual war winners while controlling for the other POIs provides the following results:

N	−2 Log Likelihood	Goodness of Fit	Model Chi-Square (significance)
22	10.580	9.231	19.918 (0.0005)
Variable	B	Std. Error	R
$\sum_{i=War} AT$ (WEI)	−3.0005	3.922	0.000
Duration	0.0216	0.023	0.000
Attrition	0.0414	0.052	0.000
Front Displacement	0.3042	2.654	0.000
Constant	3.4662	4.375	—

The classification table for this model is the same as Table 4.12, except that a single winner is wrongly predicted (i.e., in the upper right box in the table).

77. Examination of the data shows this outlier is caused by Kuwaiti military spending in the year prior to the war. The substantial increase in this spending distorts the subsequent military quality of the desert country's forces, making them appear (at least to the model) considerably more formidable than the Iraqis found them in August of 1990. There is little doubt that modification of the coding procedure for this case would improve the predictive power of several of the models in this chapter. However, in the interest of consistency, we have retained the coding outlined above.

78. Brzoska and Pearson, *Arms and Warfare,* 1994.

79. Full results of Pearson's and Rank Order tests are available in Appendix 4.

80. Bennett and Stam, "The Duration of Interstate Wars, 1816–1985," 1996, use a hazard model to predict war duration in 77 cases of interstate war from 1816–1985. Based on a careful reading of their results, a similar model for use in this chapter was rejected on the following bounds. Their model's predicted error (13 months) was excessive, and the errors were quite skewed. Furthermore, whereas the model above (which would not be useful for emulation in this study) was quite well specified (with about a dozen predictor variables and time-varying covariates), a naive model which they ran (and would be useful for emulation here) performed more poorly than did the specified model by a significant degree (four months additional mean error). Since a similar naive model would of necessity be used in any analysis in this chapter, this degree of error (or probably even larger, based on a considerably smaller sample of cases) would be unacceptable. One might note that 17 months represents the mean length of war in this data set, *even if* the Iraq-Iran war is retained. In addition, the 20 cases represented in the data set under consideration here would not provide reliable parameter estimates for the hazard model. See William Greene, *Econometric Analysis,* 3d ed. (Upper Saddle River, NJ: Prentice-Hall, 1997); and Eric Hanushek and John Jackson, *Statistical Methods for Social Scientists* (New York: Academic Press, 1977).

81. Full results of the ordinary least squares regression test is available in Appendix 4.

82. While the technical properties of the data for duration violate the assumption of normally distributed errors made by ordinary least squares because they are all positive, these results are reported because the test statistic for the model (*f-score*) is not affected. Applying a linear model to casualty data would introduce specification errors, biased coefficients and the possible prediction of negative values for the dependent variable, according to Hanushek and Jackson, *Statistical Methods for Social Scientists*, 1977; and Gary King, *Unifying Political Methodology*, (Cambridge: Cambridge University Press, 1989).

83. See note 82.

84. Complete statistics for the test of Model 4-5 are available in Appendix 4.

85. Bennett and Stam, "The Duration of Interstate Wars, 1816–1985," 1996.

86. This type of detail is available in some military models, where specific weapons are analyzed in what may best be described as "acquire and destroy" models of combat, which seek realism at the individual weapon-system level (see, for example, Fanica Gavril, "A Synthesis of Methods for Simulating Vulnerability of Armored Vehicles," in *Warfare Modeling*, ed. Bracken, Kress, and Rosenthal, 1995, 101–20; Wayne Hughes, "A Salvo Model of Warships in Missile Combat Used to Evaluate Their Staying Power," in *Warfare Modeling*, ed. Bracken, Kress, and Rosenthal, 1995, 121–44; or Martin van Dongen and Joost Kos, "The Analysis of Ship Air Defense: The Simulation Model SEAROADS," in *Warfare Modeling*, ed. Bracken, Kress, and Rosenthal, 1995, 145–63.

87. An example of this may well be the U.S. casualties during Operation Desert Storm. Of the 200 or so U.S. casualties in the second part of the Persian Gulf War, about 35 occurred due to a barracks in the rear areas being destroyed by a random SCUD missile strike. Closer to the front, American forces were more dispersed as a rule, and therefore not sleeping in large, centralized barracks due to the fear of artillery strikes (concentrated Iraqi firepower). Dunnigan, *How to Make War*, 1988; Dupuy, *Numbers, Predictions and War,* 1979; Dupuy, *Analysis of Factors That Have Influenced Outcomes of Battles and Wars,* 1983; and Michael Howard, *War in European History* (London: Oxford University Press, 1976).

4

Appendix 4
Models, Parameters, and Results: Estimating the Effects of Arms Transfers on War Using the Dynamic Force Attrition Model

Wars, Matches, and Predominant Terrain for Chapter 4 Analysis

War Dyad		Matched Dyad (Country matched)	
India-Pakistan (1971)	Rolling	Thailand (Pakistan)-Burma	Forest
Cambodia-Vietnam (1975)	Forest	Malaysia (Vietnam)-Singapore	Forest
Syria, Egypt, Jordan-Israel (1973)	Desert	NONE	
Iraq-Kuwait (1990)	Desert	Israel (Iraq)-Jordan	Desert
Syria-Israel (1982)	Desert	Libya (Syria)-Tunisia	Desert
Tanzania-Uganda (1978)	Rolling	Gabon (Tanzania)-Sierra Leone	Forest
Somalia-Ethiopia (1977)	Desert	Guinea (Somalia)-Cameroon	Forest
China-Vietnam (1979)	Forest	Bangladesh (Vietnam)-Burma	Forest
Iraq-Iran (1980)	Desert	United Arab Emirates (Iraq)-Oman	Desert
China-Vietnam (1985)	Forest	Bangladesh (Vietnam)-Burma	Forest
Israel-Egypt (1969)	Desert	Iran (Egypt)-Afghanistan	Desert

Border Wars Not Included in Chapter 4 Analysis[1]

Soviet Union-Hungary (1956)
Egypt-Israel (1956)
Arab-Israel (1967)
China-India (1962)
India-Pakistan (1965)
El Salvador-Honduras (1969)

A Dynamic Model of Force Attrition

Epstein's models for a battle are:

1. $A_g(t) = A_g(t-1)[1 - \alpha(t-1)] - DCAS(t-1) + RA_g(t),$

    where $A_g(t)$ is the attacker's ground lethality surviving at the start of
    the tth day; $DCAS_t$ is the attacking ground lethality killed by the
    defender's close air support on the tth day; RA_g are the attacker's
    reinforcements;[2] and t is the time in days. And

2. $D_g(t) = D_g(t-1) - \dfrac{\alpha(t-1)}{\rho} A_g(t-1) - ACAS(t-1) + RD_g(t),$

    where $D_g(t)$ is the attacker's ground lethality surviving at the start of
    the tth day; ρ is the attacker's ground lethality killed per defender's
    ground lethality killed (the average ground-to-ground casualty-
    exchange ratio); $ACAS_t$ is the defending ground lethality killed by
    the attacker's close air support on the tth day; and RD_g is the
    defender's reinforcements.[3] Also,

3. $\alpha(t) = \alpha_g(t)\left(1 - \dfrac{W(t)}{W_{\max}}\right),$

    where $\alpha(t)$ is the attacker's ground-to-ground lethality attrition rate
    per day (no airpower case), $0 \le \alpha(t) \le 1$; $\alpha_g(t)$ is the attacker's
    ground prosecution rate per day, $0 \le \alpha_g(t) \le 1$; $W(t)$ is the
    defender's rate of withdrawal in kilometers per day; and $W_{\max}$ is the
    defender's maximum rate of withdrawal.

    If $\alpha_d(t-1) \le \alpha_{dT}$, where α_d is the defender's total ground-le-
    thality attrition rate per day (air and ground induced), $0 \le \alpha_d(t) \le 1$;

and α_{dT} is the defender's threshold attrition rate, the value beyond which withdrawal begins, $0 \le \alpha_{dT} \le 1$; then

4. $$W(t) = \begin{cases} 0 \\ W(t-1) + \left(\dfrac{(W_{\max} - W(t-1))}{1 - \alpha_{dT}} \right) (\alpha_d(t-1) - \alpha_{dT}). \end{cases}$$

However, if $\alpha_d(t-1) > \alpha_{dT}$ and

5. $$\alpha_d(t) = \frac{D_g(t) - \left[D_g(t+1) - RD_g(t+1) \right]}{D_g(t)}.$$

$W(t)$ will initially be set to 0.

On the attacker's side,

6. $$\alpha_g(t) = \alpha_g(t-1) - \left(\frac{\alpha_{aT} - \alpha_g(t-1)}{\alpha_{aT}} \right) (\alpha_a(t-1) - \alpha_{aT}),$$

where α_{aT} is the attacker's threshold, or equilibrium, attrition rate; the value of $\alpha_a(t)$ the attacker seeks to achieve and sustain, $0 \le \alpha_{aT} \le 1$; and $\alpha_a(t)$ is the attacker's total ground-lethality attrition rate per day (air and ground induced), $0 \le \alpha_a(t) \le 1$; with

7. $$\alpha_a(t) = \frac{A_g(t) - \left[A_g(t+1) - RA_g(t+1) \right]}{A_g(t)}.$$

$\alpha_g(t)$ is set to some initial value so that $\alpha_g(1) < \alpha_{aT}$.

Likewise, the terms for the air components of the battle are for the defender:[4]

8. $$DCAS(t) = \frac{L}{V} D_a(1)(1 - \alpha_{da})^{Sd(t-1)} K_a \left[\frac{1 - (1 - \alpha_{da})^{Sd+1}}{\alpha_{da}} - 1 \right],$$

where L are the lethality points per division equivalent; V is the number of armored fighting vehicles per division equivalent; $D_a(t)$ is the defender's CAS aircraft surviving at the start of the tth day; α_{da} is the defender's CAS aircraft attrition rate per sortie, $0 \le \alpha_{da} \le 1$; S_d is the defender's CAS daily sortie rate; and K_d is the attacker's armored fighting vehicles killed per defender CAS sortie. For the attacker:

9. $$ACAS(t) = \frac{L}{V} A_a(1)(1 - \alpha_{aa})^{Sa(t-1)} K_a \left[\frac{1 - (1 - \alpha_{aa})^{Sa+1}}{\alpha_{aa}} - 1 \right],$$

where $A_a(t)$ is the attacker's CAS aircraft surviving at the start of the tth day; α_{aa} is the attacker's CAS aircraft attrition rate per sortie,

$0 \leq \alpha_{aa} \leq 1$; S_a is the attacker's CAS daily sortie rate; and K_a is the defender's armored fighting vehicles killed per attacker CAS sortie.

In order to run the data in this model, the following values were assigned (as suggested by Epstein 1990):

Variable	Assumed Value	Variable	Assumed Value
ρ	1.7^5	S_a and S_d	1^6
α_{aT}	0.06	K_d	7.5
W_{max}	20	K_a	7.5
α_{da}	0.05	$W(1)$	0
α_{aa}	0.05	$\alpha_g(1)$	0.03

Also, the following alteration was made:

- L/V is replaced in the model with a constant reflecting the amount of ground lethality killed by airpower on each day. Instead of basing this value on a high number of vehicles in an armored division, as is Epstein's estimation based on the European NATO-Warsaw Pact confrontation, we base it on the equivalent value in WEI "points" vis the total force in WEIs. In this estimation, since a single (average value) armored vehicle is worth approximately 7.5 WEI (.75 × 10.0) on the baseline "rolling terrain," this is the amount of loss caused per sortie by both attacker and defender aircraft (based on the above values for sortie rate and kills per sortie).

Full Results of POI Difference of Means Tests
(Hypothesis 4-1)

Data and Analysis

Comparison of Means of Pre- and Posttransfer
Predicted Outcomes (Days)

Dyad	Year	Pretransfer Prediction	Posttransfer Prediction
India-Pakistan	1971	46	8
Vietnam-Cambodia	1975	40	19
Arab-Israel	1973	40	38
Iraq-Kuwait	1990	40	40
Israel-Syria	1982	8	10
Uganda-Tanzania	1978	39	37
Somalia-Ethiopia	1977	41	6
China-Vietnam	1979	4	4
Israel-Egypt	1969	19	30
Iraq-Iran	1980	39	38
China-Vietnam	1985	7	7
Thailand-Burma*	1971	20	13
Malaysia-Singapore*	1975	40	40
Afghanistan-Iran*	1973	36	3
Israel-Jordan*	1990	2	2
Libya-Tunisia*	1982	3	3
Gabon-Cameroon*	1978	48	30
Guinea-Sierra Leone*	1977	18	18
Bangladesh-Burma*	1979	67	38
United Arab Emirates-Oman*	1980	40	38
Bangladesh-Burma*	1985	47	39
	MEAN	29.2381	21.9524
	t-value		−2.13
	2-tail significance		0.046

*   Indicates control dyads of matched pairs.

Comparison of Means of Pre- and Posttransfer
Predicted Outcomes (Attacker Attrition)

Dyad	Year	Pretransfer Prediction	Posttransfer Prediction
India-Pakistan	1971	90	27
Vietnam-Cambodia	1975	90	49
Arab-Israel	1973	90	90
Iraq-Kuwait	1990	90	90
Israel-Syria	1982	16	21
Uganda-Tanzania	1978	90	90
Somalia-Ethiopia	1977	90	100
China-Vietnam	1979	10	11
Israel-Egypt	1969	38	82
Iraq-Iran	1980	90	90
China-Vietnam	1985	16	17
Thailand-Burma[*]	1971	51	36
Malaysia-Singapore[*]	1975	90	90
Afghanistan-Iran[*]	1973	90	33
Israel-Jordan[*]	1990	4	4
Libya-Tunisia[*]	1982	6	6
Gabon-Cameroon[*]	1978	72	57
Guinea-Sierra Leone[*]	1977	41	41
Bangladesh-Burma[*]	1979	90	73
United Arab Emirates-Oman[*]	1980	68	67
Bangladesh-Burma[*]	1985	79	71
	Mean	61.9524	54.0476
	t-value		−1.59
	2-tail significance		0.127

[*] Indicates control dyads of matched pairs.

Comparison of Means of Pre- and Posttransfer
Predicted Outcomes (Winners)

Dyad	Year	Pretransfer Prediction	Posttransfer Prediction
India-Pakistan	1971	Attacker	Attacker
Vietnam-Cambodia	1975	Defender	Attacker
Arab-Israel	1973	Defender	Defender
Iraq-Kuwait	1990	Defender	Defender
Israel-Syria	1982	Attacker	Attacker
Uganda-Tanzania	1978	Defender	Defender
Somalia-Ethiopia	1977	Defender	Defender
China-Vietnam	1979	Attacker	Attacker
Israel-Egypt	1969	Attacker	Attacker
Iraq-Iran	1980	Defender	Defender
China-Vietnam	1985	Attacker	Attacker
Thailand-Burma*	1971	Attacker	Attacker
Malaysia-Singapore*	1975	Defender	Defender
Afghanistan-Iran*	1973	Defender	Attacker
Israel-Jordan*	1990	Attacker	Attacker
Libya-Tunisia*	1982	Attacker	Attacker
Gabon-Cameroon*	1978	Attacker	Attacker
Guinea-Sierra Leone*	1977	Attacker	Attacker
Bangladesh-Burma*	1979	Attacker	Attacker
United Arab Emirates-Oman*	1980	Attacker	Attacker
Bangladesh-Burma*	1985	Attacker	Attacker
	Mean	0.2381	0.4286
	t-value		1.45
	2-tail significance		0.162

*   Indicates control dyads of matched pairs.

Comparison of Means of Pre- and Posttransfer
Predicted Outcomes (Front Displacement)

Dyad	Year	Pretransfer Prediction	Posttransfer Prediction
India-Pakistan	1971	368.3	75.48
Vietnam-Cambodia	1975	0	111.96
Arab-Israel	1973	0	0
Iraq-Kuwait	1990	0	0
Israel-Syria	1982	87.4	107.72
Uganda-Tanzania	1978	0	0
Somalia-Ethiopia	1977	0	0
China-Vietnam	1979	28.33	26.24
Israel-Egypt	1969	197.27	218.69
Iraq-Iran	1980	0	0
China-Vietnam	1985	62.36	60.87
Thailand-Burma[*]	1971	171.69	118.1
Malaysia-Singapore[*]	1975	0	0
Afghanistan-Iran[*]	1973	0.07	27.84
Israel-Jordan[*]	1990	12.64	12.39
Libya-Tunisia[*]	1982	24.28	24.28
Gabon-Cameroon[*]	1978	506.7	289.12
Guinea-Sierra Leone[*]	1977	161.03	161.32
Bangladesh-Burma[*]	1979	598.55	330.67
United Arab Emirates-Oman[*]	1980	397.03	371.87
Bangladesh-Burma[*]	1985	435.03	365.98
	Mean	145.2705	109.6466
	t-value		−1.63
	2-tail significance		0.119

[*]   Indicates control dyads of matched pairs.

Results of Tests of Hypotheses 4-2 and 4-3

The Effect of Arms Transfers on War Involvement when Difference in Pre- and Post-Arms Transfer WEI Ratio Controlled

N	−2 Log Likelihood	Goodness of Fit	Model Chi-Square (significance)	
42	40.020	44.048	18.109(.0001)	
Variable	B	Std. Error	Significance	R
$\sum_{i=3} AT$ (WEI)	0.0002	9.153	0.012	0.268
DIFFRATIO$_t$ (Difference in Pre- and Post-Aggregate WEI Values)	0.2712	0.491	0.580	0.000
Constant	−1.1249	0.480	0.019	----

Classification Table for Arms Transfers on War Involvement when Difference in Pre- and Post-Arms Transfer WEI Ratio Controlled

	Predicted		
Observed	Loser	Winner	Percent Correct
Loser	19	1	95.0
Winner	7	15	68.18
		Overall	80.95

The Effect of Arms Transfers on War Involvement when
Difference in Pre- and Post-Arms Transfer POI Controlled

Initial Log Likelihood Function

-2 Log Likelihood 58.129089
* Constant is included in the model.

Estimation terminated at iteration number 6 because
Log Likelihood decreased by less than .01 percent.

–2 Log Likelihood	34.897
Goodness of Fit	31.691

	Chi-Square	df	Significance
Model Chi-Square	23.232	5	.0003
Improvement	23.232	5	.0003

Variables in the Equation

Variable	B	S.E.	Wald	df	Sig	R	Exp(B)
Pctattri	–0.0439	0.0323	1.8452	1	0.1743	0.0000	0.9570
Pctdays	0.0073	0.0154	0.2235	1	0.6364	0.0000	1.0073
Pterrito	0.0134	0.0067	4.0123	1	0.0452	0.1861	1.0135
Win	0.9414	2.3927	0.1548	1	0.6940	0.0000	2.5637
Atval	0.0004	0.0002	6.8529	1	0.0088	0.2889	1.0004
Constant	–1.2594	0.5827	4.6716	1	0.0307		

Classification Table for Arms Transfers on War Involvement
when Difference in Pre- and Post-Arms Transfer POI Controlled

	Predicted		
Observed	Loser	Winner	Percent Correct
Loser	17	3	85.00
Winner	6	16	72.73
		Overall	78.57

Test of Hypothesis 4-5, Model 4
Relationship between War Arms Transfers and Duration in Nation-Months

Pearson's Correlation Coefficients

	Natmonth	Ratio1	Warat
Natmonth	1.0000	–0.1390	0.3000
	(20)	(20)	(20)
	P = .	P = 0.559	P = 0.199
Ratio1	–0.1390	1.0000	–0.0762
	(20)	(42)	(22)
	P = 0.559	P = .	P = 0.736
Warat	0.3000	–0.0762	1.0000
	(20)	(22)	(22)
	P = 0.199	P = 0.736	P = .

(Coefficient / (Cases) / 2-tailed Significance)
" . " is printed if a coefficient cannot be computed

Spearman Correlation Coefficients

Ratio1	.0000	
	N(20)	
	Sig1.000	
Warat	.2937	.0695
	N(20)	N(22)
	Sig .209	Sig .759
	Natmonth	Ratio1

(Coefficient / (Cases) / 2-tailed Significance)
" . " is printed if a coefficient cannot be computed

Ordinary Least Squares Regression Test of War Arms Transfers on Duration with Ratio (WEI) Controlled

Multiple R	.32850
R Square	.10791
Adjusted R Square	.00296
Standard Error	13.34305

Analysis of Variance

	DF	Sum of Squares	Mean Square
Regression	2	366.10903	183.05451
Residual	17	3026.62897	178.03700

F = 1.02818   Signif F = .3789

Variables in the Equation

Variable	B	SE B	Beta	T	Sig T
Ratio1	−0.822679	1.407599	−0.133905	−0.584	0.5666
Warat	0.002721	0.002095	0.297668	1.299	0.2112
(Constant)	7.710043	4.747425		1.624	0.1228

**Ordinary Least Squares Regression Test of War
Arms Transfers on Duration in Nation-Months
with Predicted Outcome Indicators Held Constant**

Multiple R	0.31708
R Square	0.10054
Adjusted R Square	−.22069
Standard Error	14.76392

Analysis of Variance

	DF	Sum of Squares	Mean Square
Regression	5	341.11041	68.22208
Residual	14	3051.62759	217.97340

F = .31298   Signif F = .8970

Variables in the Equation

Variable	B	SE B	Beta	T	Sig T
Warat	0.002812	0.002602	0.307592	1.081	0.2981
Attwat	−0.048367	0.254341	−0.115791	−0.190	0.8519
Dayswat	0.025000	0.355431	0.046777	0.070	0.9449
Predwin	−0.315246	28.088151	−0.012102	−0.011	0.9912
Terwat	0.009324	0.114350	0.064847	0.082	0.9362
(Constant)	2.803636	29.898625		0.094	0.9266

The Effect of Arms Transfers on Actual Casualty Rates

Pearson Correlation Coefficients: War Arms Transfers, Prewar Military Capabilities Ratio (WEI Values), and Battle Deaths

	Bdeaths	*Ratio1*	*Warat*
Bdeaths	1.0000	0.1882	0.0532
	(20)	(20)	(20)
	P = .	P = 0.427	P = 0.824
Ratio1	0.1882	1.0000	–0.0762
	(20)	(42)	(22)
	P = 0.427	P = .	P = 0.736
Warat	0.0532	–0.0762	1.0000
	(20)	(22)	(22)
	P = 0.824	P = 0.736	P = .

(Coefficient / (Cases) / 2-tailed Significance)
" . " is printed if a coefficient cannot be computed

Spearman Correlation Coefficients: War Arms Transfers, Prewar Military Capabilities Ratio (WEI Values), and Battle Deaths

Ratio1	–0.1744	
	N(20)	
	Sig 0.462	
Warat	0.1880	0.0695
	N(20)	N(22)
	Sig 0.427	Sig 0.759
	Bdeaths	Ratio1

(Coefficient / (Cases) / 2-tailed Significance)
" . " is printed if a coefficient cannot be computed

Model 4-5 with Ratio (WEI) Controlled

Multiple R	0.19646
R Square	0.03860
Adjusted R Square	–0.07451
Standard Error	4076.23928

Analysis of Variance

	DF	Sum of Squares	Mean Square
Regression	2	11339696.28614	5669848.14307
Residual	17	282467352.91386	16615726.64199

F = .34123   Signif F = .7157

Variables in the Equation

Variable	B	SE B	Beta	T	Sig T
Ratio1	341.954028	430.014830	0.189137	0.795	0.4375
Warat	0.151975	0.639903	0.056487	0.237	0.8151
(Constant)	2978.893717	1450.315927		2.054	0.0557

Pearson Correlation Coefficients: War Arms Transfers, Predicted Attrition Rate, Battle Deaths, Predicted Front Displacement, Predicted Winners, Predicted Duration

	ATT WAT	BDEA THS	DAYS WAT	PRED WIN	TER WAT	WAR AT
Attwat	1.0000	0.1355	−0.1752	0.6885	0.4454	0.1422
	(42)	(20)	(42)	(22)	(42)	(22)
	P = .	P = 0.569	P = 0.267	P = 0.000	P = 0.003	P = 0.528
Bdeaths	0.1355	1.0000	−0.1416	−0.1252	−0.1203	0.0532
	(20)	(20)	(20)	(20)	(20)	(20)
	P = 0.569	P = .	P = 0.552	P = 0.599	P = 0.614	P = 0.824
Dayswat	−0.1752	−0.1416	1.0000	0.2842	−0.5104	−0.2521
	(42)	(20)	(42)	(22)	(42)	(22)
	P = 0.267	P = 0.552	P = .	P = 0.200	P = 0.001	P = 0.258
Predwin	0.6885	−0.1252	0.2842	1.0000	0.6326	−0.1567
	(22)	(20)	(22)	(22)	(22)	(22)
	P = 0.000	P = 0.599	P = 0.200	P = .	P = 0.002	P = 0.486
Terwat	0.4454	−0.1203	−0.5104	0.6326	1.0000	0.0069
	(42)	(20)	(42)	(22)	(42)	(22)
	P = 0.003	P = 0.614	P = 0.001	P = 0.002	P = .	P = 0.976
Warat	0.1422	0.0532	−0.2521	−0.1567	0.0069	1.0000
	(22)	(20)	(22)	(22)	(22)	(22)
	P = 0.528	P = 0.824	P = 0.258	P = 0.486	P = 0976	P = .

(Coefficient / (Cases) / 2-tailed Significance)
" . " is printed if a coefficient cannot be computed

**Spearman Correlation Coefficients: War Arms Transfers,
Predicted Attrition Rate, Battle Deaths, Predicted Front Displacement,
Predicted Winners, Predicted Duration**

Bdeaths	−0.1025				
	N(20)				
	Sig 0.667				
Dayswat	−0.3100	0.0672			
	N(42)	N(20)			
	Sig 0.046	Sig 0.778			
Predwin	0.7642	−0.2177	0.2079		
	N(22)	N(20)	N(22)		
	Sig 0.000	Sig 0.357	Sig 0.353		
Terwat	0.8040	−0.2133	−0.5385	0.7223	
	N(42)	N(20)	N(42)	N(22)	
	Sig 0.000	Sig 0.366	Sig 0.000	Sig 0.000	
Warat	0.0293	0.1880	0.0226	0.1361	0.0421
	N(22)	N(20)	N(22)	N(22)	N(22)
	Sig 0.897	Sig 0.427	Sig 0.920	Sig 0.546	Sig 0.852
	Attwat	Bdeaths	Dayswat	Predwin	Terwat

(Coefficient / (Cases) / 2-tailed Significance)
" . " is printed if a coefficient cannot be computed

Model 4-5 with POIs Controlled

Multiple R	.44368
R Square	.19686
Adjusted R Square	−.08998
Standard Error	4105.47918

Analysis of Variance			
	DF	*Sum of Squares*	*Mean Square*
Regression	5	57837618.71458	11567523.74292
Residual	14	235969430.48542	16854959.32039

F = .68630   Signif F = .6417

Variables in the Equation

Variable	B	SE B	Beta	T	Sig T
Attwat	90.489925	70.725831	0.736165	1.279	0.2215
Dayswat	6.056761	98.836576	0.038511	0.061	0.9520
Warat	0.362036	0.723551	0.134565	0.500	0.6246
Predwin	−5564.770086	7810.615438	−0.725941	−0.712	0.4879
Terwat	−1.583029	31.797759	−0.037413	−0.050	0.9610
(Constant)	12546.454048	8314.063203		1.509	0.1535

Notes

1. Wars not included because of absence of reliable data.

2. Because there is no basis upon which to estimate the reinforcements of either side in the test or control cases for the data used in Chapter 3, we will omit Ra_g from the analysis. To do so does not effect the integrity of the model as presented here; indeed, Epstein's first presentation of the model did not include this variable (Epstein 1985).

3. The defender's reinforcements are also not estimated for the same reasons as above.

4. Terms for air reinforcements have been omitted for the same reasons as above.

5. Epstein's use of this exchange rate revises his 1985 estimate of 1.5, giving a greater advantage to the defender. This revision is based on the recently available historical studies by Helmbold and Khan (1986), McQuie (1988a), and McQuie (1988b). To observe how changes in this value affect force planning based on the Epstein model, see Epstein 1990, Appendix D.

6. Values for sorties per day for each side are not taken from Epstein. They are approximation based on the logic that most countries could not maintain the sortie rate of either the Warsaw Pact (two per day) or NATO (three per day) during an all-out confrontation.

5

Conclusion: Weapons for Future Peace, Weapons for Future War

To the moral man, a discussion of the function of violence should be both realistic and unenthusiastic. It can be established, we believe, that the facts of social and political life are still such as to make the use of violence necessary, under certain limited circumstances, in order to achieve an objective the non-achievement of which would constitute a greater evil than that involved in such use of violence. But the need for violence is just as regrettable as the existence of those facts which necessitate it, and any exaltation or joy in its use is a definite perversity. The whole subject is a delicate one which the moralist and the political scientist must approach with clear concepts, exact definitions, and close reasoning.

Martin Hillenbrand, *Power and Morals*[1]

The institution of human slavery was created at the dawn of the human race, and many once felt it to be an elementary fact of existence. Yet between 1788 and 1888 the institution was substantially abolished . . . and this demise seems to be permanent. Similarly the venerable institutions of human sacrifice, infanticide and dueling seem also to have died out or been eliminated. It could be argued that war, at least war in the developed world, is following a similar trajectory.

John Mueller, *British Journal of Political Science*[2]

What Has Gone Before

The previous chapters of this book examined basic questions about a phenomenon—war—that, until very recently, appeared to dominate the study of politics. The statements by Hillenbrand and Mueller indicate that questions related to this topic deserve "close reasoning," and that there is hope that this "institution" (war) can be eradicated.[3] Unfortunately, without careful examination of indicators of war-prone behavior, such an eradication will not occur. It is to this task that we have turned in this book. Specifically, we examined questions related to the impact of weapons sales on war outbreak, involvement, termination, and other outcomes. These examinations brought forth new data analyses of the relationship between arms transfers at the global level, among supplier and recipient states, and by using military science techniques to examine bordering dyads.

Some general relationships can be discerned from the previous chapters. Each chapter's data analysis supported in varying degrees the basic contention that arms transfers are related to war involvement. However, when it came to war outcomes, we found very little evidence of any relationship with arms transfers. For many reasons, arms transfers may not be useful as predictors of such war outcomes as who wins, how bloody the war will be, and how long the war will last. It is tempting to be neither profound nor original and conclude that weapons really do not make, or determine the outcomes, of war.

Humans do, however. Humans also make the decisions as to whether to sell weapons or refrain from such sales, and such sales do predate almost every war in our data.[4] As noted by Seldes, it used to be that

> almost all governments kept up the pretense of disinterestedness in the armament business. Publicly it was a policy of *laissez-faire*. Gunmaking was a business, like any other, and there were no restrictions. No Prime Minister or President would admit he encouraged or directed the sale of munitions.[5]

Unfortunately, Wolpin and Hartung show that this is not the case today.[6] In the modern global weapons trade, even more so than when Seldes penned *Iron, Blood and Profits,* governments cooperate with the munitions interests; they name favored nations to arm and float loans for armaments. When an American President specifies which party in a Central or South American revolution is righteous and which the villain, and gives his sanction to the shipments of arms to the one while prohibiting them to the other, that President may be said to participate in the private business of the rifle dealers.[7]

Indeed Hartung provides evidence that the United States post–Cold War arms sales policy often had U.S. military, State Department, and other executive branch officials actively involved in marketing, demonstrating, and otherwise pursuing buyers. During this period, the U.S. became the world's largest weapons supplier.[8]

This study finds that such sales contribute to states' involvement in wars. Hillenbrand is right; after careful consideration and after the exhaustion of conflict resolution by other means, some wars may be worth fighting. If modern governments are selling munitions and weapons to other countries because they are fighting mutually agreed upon, "worthwhile" wars, then the weapons trade is perhaps commendable. However, we may have doubts that this is the case. As noted in Chapter 3, the U.S. government notifies the American people that its weapons sales are not for war-fighting purposes, do not "threaten the stability" of the region into which they are transferred, and do not alter the "military balance" between the recipient state and (presumably) its enemies. The findings in Chapter 4 indicate that the focus on the "military balance" may introduce distortions into the perception of the likely impact of a given arms transfer. The analysis there showed that yearly arms imports have no distinguishable effect on the military balance between two states. Yet knowing the level of arms transfers and controlling for the military balance of the states prior to the war (rather than the change in military balance due to arms transfers) allows us to predict war involvement of the states with considerable success—even with a very simple model.

According to Hartung and others, weapons sales are more likely to be *actively promoted* by leaders in the major arms supplying states, many of which are "peaceful" democracies that seem to have among themselves reached the status of "war abolition" that Mueller describes. Yet the arms manufacturers, or merchants of death in the language of Engelbretch and Hanighen, export their goods to other states who have, for the most part, not bought into the abolition of war (nor the "democratic peace").[9] They do so in order to maintain domestic political support—sometimes at the expense of the lethal repression of dissident groups within the recipients' societies—and to meet the economic and security interests of foreign policy.[10] It appears, then, that the "abolition of war" is quite different from the abolition of slavery, where at least the object of enslavement was freed (and eventually promised 40 acres (Gen. Sherman's promise, 1865), a mule, and a plough). In the "war abolitionist movement" in the modern world, the objects of war are not to be "beaten into ploughshares," but rather continually made more effective, deadly, and destructive and sold to those who have not abolished war as the "continuation of political intercourse with the intermixing of other means."[11] It is doubtful that Clausewitz would be comfortable with the intermixture of war with weapons of mass destruction and today's advanced conventional weapons technology. William Tecumseh Sherman, given his definition of war, would.[12]

It seems that despite the growing weight of evidence that arms transfers increase the likelihood of a state becoming involved in a war, governments will not control the spread of weapons across the globe.[13] Given these realities, it is useful to examine some variations on the perceived future of warfare and what the renewed upswing in the global weapons market portends.[14]

The Future of War and the Global Arms Market

A review of prominent forecasts of military analysts, international relations scholars, and foreign policy elites allows the construction of a typology of future war (see Table 5.1). This section provides a description of the five views of future war, keeping in mind Hillenbrand's admonition against war's utilization as policy for transient reasons and Mueller's abolitionist hopes, and analyzes the role of the weapons trade in their occurrence.

No Intensity Future Warfare

Scholars such as Arnold and Arquilla and Ronfeldt articulate "no intensity" concepts of future war based on nonviolent and nonlethal attacks on societal economic, technological, or informational infrastructures.[15] Attacking a society's economic "center of gravity," for instance, might entail sabotage of the central apparatus for controlling a state's stock market, causing financial chaos, and perhaps irrecoverable loss of assets within the state. Such conceptions of future war do not impact upon what we traditionally define as warfare; neither are they waged with what are traditionally considered weapons. However, it is a matter of common sense that virtually any object can be used as a weapon, and when we consider the history of warfare we quickly realize that "no intensity" mechanisms have always been present.[16] Just because deadly force is not used does not mean that the underlying basis of the state (or other organized grouping) cannot be attacked. By doing so, the attacker may undermine the overall power of the enemy no less than if he had used laser-guided munitions to take out the "command structure" and "power grid" in a manner akin to the Warden/Horner plan of attack against Iraq in 1991.[17]

Table 5.1: Typology of Future War

No Intensity	*Low Intensity*	*Medium Intensity*	*High Intensity*	*Very High Intensity*
Economic, technological and information warfare	Intrastate wars, wars of "failed states," low-intensity conflicts	"Medium-sized regional contingencies," regional warfare, wars of civilizations	Warfare between great powers with high-tech conventional weapons	Warfare with weapons of mass destruction usage

Low-Intensity Future Warfare

Low-intensity warfare, or insurgent war, "is the culmination of what many militarily inferior forces have learned to do to succeed against stronger foes ... it is 'warfare on the cheap.' "[18] Insurgencies typically serve the political purpose of overthrowing a government and replacing it with one more compatible with dissatisfied groups' interests, or of repelling militarily more powerful invaders. Bozeman notes that these wars are central where ideas clash and where the primacy of the state, either to govern or as a means to govern, is challenged.[19] Creveld asserts that these wars, which are often ingrained in the culture of the peoples undertaking them, are at once older than the state and likely to outlive the state as a useful concept in world affairs.[20] The low-intensity nature of these types of war is necessitated by the relative weakness of the insurgent forces, who often have more manpower than weapons (and almost always more so than major weapons). Such wars are fought with "small arms" that are relatively unregulated in the global marketplace. However, as noted by Keegan and numerous others, these "small arms" pack a big punch by virtue of the fact that on a *yearly* basis they kill more people than all of the nuclear weapons built since the beginning of August 1945.[21] These "small arms" are available from places like Interarms, one of 250 or so supplier corporations listed by *Jane's Infantry Weapons 1994–95*, and they account for roughly $20 billion a year in the global weapons trade.[22]

Medium-Intensity Future Warfare

Medium-intensity warfare in the future will lie, as it does today, somewhere between low-intensity conflicts and great power wars. The Persian Gulf War (1991) between the United Nations coalition and Iraq was a medium-intensity conflict, as was the Iraq-Iran war (1980–1988). These wars are likely to take place, according to Snow, as part of "militarized states' " struggle for regional hegemony and because of the renewed struggle for resources in Third World areas due to the withdrawal of superpower aid. Weapons, sometimes highly advanced ones but usually those that become great power surplus, are likely to flow to these regions due to the renewed competition between the great powers for arms markets.[23] The great powers may become less able or willing to restrain such regional conflicts, and when they intervene, it will be on the terms outlined by Rice at the conclusion of the Persian Gulf War, "the object will be quick action and few casualties."[24] By assuring that those great powers that develop global reach and global interests (currently only the U.S.) in the future retain a technological superiority over regional aggressors, these terms can be achieved even as the regional challengers inevitably become more sophisticated themselves.

High-Intensity Future Warfare

Some recent examinations of the future espouse that a war may soon occur between the U.S. and "a future peer."[25] These forecasts operate under the assumption that U.S. global hegemony is unnatural, and that eventually a challenger will appear. This challenger may be able to create military capabilities that are equal to the U.S. across a range of technologies, or in various "niches" that are extremely high value. A war between such competitors, if it does not reach a "very high intensity" level (see below), will involve high-tech conventional weapons usage on a massive scale, "deep" battlefields (strikes occurring in tactical, operational, theater, and strategic modes), and holistic concepts of operations that include the "no intensity" warfare options discussed above. It would be a war controlled by military science to the effect that, when there is no longer any doubt about which side is stronger, the war would of necessity end, since neither side could win a very high-intensity conflict.[26]

Very High-Intensity Future Warfare

Interestingly enough, very high-intensity warfare is the most difficult to discuss in terms of future wars, and it appears that no one does so to any great length.[27] However, the literature from the Cold War is familiar to most analysts, and it seems safe to say that the "old" theories of assured destruction, counterforce versus countervalance targeting, limited nuclear wars, escalation, and extended deterrence still apply.[28] Warfare at this level would require weapons of mass destruction. However, these are themselves tightly controlled, although the technologies used to produce them are less so. The more interesting currents in this literature have to do with combinations of actors of the low- and medium-intensity wars below with acquisition of weapons of mass destruction and terrorist-type employment, and counterproliferation.[29] In the latter, one state may, by use of military force, "surgically" eliminate the weapons of mass destruction capability of another. Counterproliferation efforts may also consist of deploying antiweapons of mass destruction defensive weapons. Finally it is yet to be seen what effect the nuclearization of the India/Pakistan rivalry will have. This issue could be particularly salient given the fact that these countries are responsible for two of the wars studied in Chapter 4, and their history of enmity and proximity make any attempt to draw comparisons with the Cold War standoff between the U.S. and Soviet Union moot. When one also examines the traditional suppliers of these two states, Russia and China, and their own political and economic rivalry, one is not optimistic about the future consequences of their leap across the nuclear threshold and subsequent race to weaponization.

The above views of the future have some, but limited, utility in forecasting the arms markets of the future. At the no-intensity end of the spectrum, traditional weapons are not important, but technology transfer is. With low-intensity options, "small arms" or "light weapons" are primary. At the medium- and high-intensity levels, major conventional weapons and advanced weapon systems may be marketed. Finally, at the very high end, weapons of mass destruction and weapons designed to defeat them, or mitigate their effects, are essential. The trouble is to determine the probability of any of the above becoming the predominant form of warfare in the future or, conversely, to analyze how combinations of the above will create a synergistic arms market then.

This question is not as simple as it might appear. Many would argue that the least intense forms of warfare will be the most likely, while the very high-intensity wars are unlikely. However, the evidence surrounding an "all out", no-intensity future war is relatively small in the unclassified literature. Simply put, we do not have a good grasp on what types of information warfare operations, for instance, are being covertly waged by current—much less future—governments. If we do not have a grasp on this, then it is difficult to encourage decision makers to create policies that would reflect a rational allocation of resources to meet such challenges. The latter topic is important enough to warrant separate discussion and provides a fitting last argument for a study that found that increased weapons transfers are linked to war involvement.

Building on the Findings: Arms Control and the Relationship between Arms Transfers and War Involvement in the Future

The end of the Persian Gulf War saw first President Bush, and then President Clinton, determined to dismantle the Iraqi weapons of mass destruction programs. Almost a decade later, the U.S. continually confronts the possibility of military action in order to counter Iraqi proliferation. In the intervening years, efforts at weapons control along each range of conflict denoted above were attempted. In some cases, as with the attempts to control "small arms," the efforts have barely gotten under way, with virtually no interest shown by governments.[30] In others, such as those surrounding the proliferation of weapons of mass destruction, substantial intergovernmental, diplomatic, economic, and military policies have been developed and pursued up to the point of the recent U.S. and British counterproliferation operations against Iraq.[31] Finally, with respect to the types of technology that would seemingly be used as "weapons" with regard to the no-intensity option, controls have even been relaxed.[32]

While the statements above summarize succinctly the state of affairs for the no-intensity, low-intensity, and very high-intensity type future wars, most arms control efforts at the medium- and high-intensity levels of future war have been undertaken with suspect commitment by the states and organizations involved. The results are predictably mixed. A major supply-side conventional weapons control cartel was discussed by the United States, Russia, China, Great Britain, and France shortly after the conclusion of the Persian Gulf War and failed. The Coordinating Committee for Multilateral Export Controls (COCOM) was dismantled and rose again as the Wassenaar Arrangement on Export Controls for Conventional Arms and Dual-Use Goods and Technologies.[33] However, there are serious questions as to whether the Wassenaar Arrangement, with its 33 members, will or can be effective at controlling the trade in conventional weapons and dual-use technologies.[34] The United Nations established a Register of Conventional Weapons where states could voluntarily submit information concerning sales and purchases.[35] However, according to the most recent report, only about 25 percent of all reports given by the states can be verified completely through the UN registration process. Furthermore, the figures given by the UN do not always correspond to the figures given by ACDA or SIPRI.

It is entirely logical that a consideration of the record of arms control efforts in the post–Cold War period contributes to the prognostications of military historians, defense analysts, and political scientists concerning the future of warfare. Simply put, failures to control dangerous weapons will, in the minds of these scholars and analysts, lead to war in the future. It was the examination of whether this relationship indeed existed—whether our common sense could be trusted—that sparked the initial interest in this study. The findings in Chapters 2, 3, and 4 indicate that, with some reservations,[36] this relationship does exists.

While further research is needed to confirm the relationships established in this study, there are some recommendations that can be made. First, *additional* political, economic, and ideational resources need to be devoted to arms control. With the end of the Cold War, this has been reversed in the United States, where the government body established to be the primary proponent of arms control, the Arms Control and Disarmament Agency, is being dismantled by Congress. Second, the United States and like-minded countries (if there are any) need to *bring more resources to bear* within the Wassenaar Arrangement and United Nations in order to convince other members of the efficacy of controlling the trade in major conventional weapons. Admittedly, this will be impossible to do while the United States is the world's largest exporter of such weapons. Finally, the role of independent research organizations, who study the weapons trade, provide data concerning major conventional and light weapons sales, and attempt to influence governments to restrain the weapons trade, should be emphasized. Because governments are prone to the influences of various constituencies (e.g., those who profit from the weapons trade as well as those who seek its control), the voice of those who believe, and have scientific evidence to back up their beliefs, should be loud, clear, and persistent. If it isn't, then governments will hear other voices, and on the battlefields of the future, weapons will speak:

We are the guns, and your masters! Saw ye our flashes
Heard ye the scream of our shells in the night, and the
   shuddering crashes?

Saw ye our work by the roadside, the shrouded things lying,
Moaning to God that He made them—the maimed and the
   dying?

Husbands or sons,
Fathers or lovers, we break them. We are the guns![37]

Notes

1. Martin Hillenbrand, *Power and Morals* (New York: Columbia University Press, 1949).

2. John Mueller, "Changing Attitudes to War: The Impact of the First World War," *British Journal of Political Science* 21 (1991): 25–50.

3. Indeed, the Carnegie Commission has declared that 1997 was the most peaceful year in recent history. During that year, there were no interstate wars. Carnegie Commission on Preventing Deadly Conflict, *Preventing Deadly Conflict: Final Report* (New York: Carnegie Corporation of New York, December 1997).

4. Michael Brzoska and Frederic Pearson, *Arms and Warfare: Escalation, De-escalation, and Negotiations* (Columbia: University of South Carolina Press, 1994) make a similar observation.

5. George Seldes, *Iron, Blood and Profits: An Exposure of the World-Wide Munitions Racket* (New York: Harper and Brothers, 1934), 261.

6. Miles Wolpin, *America Insecure: Arms Transfers, Global Interventionism, and the Erosion of National Security* (London: McFarland and Company, 1991); William Hartung, *And Weapons for All: How America's Multibillion-Dollar Arms Trade Warps Our Foreign Policy and Subverts Democracy at Home* (New York: Harper Collins, 1994).

7. Seldes, *Iron, Blood and Profits*, 1934, 261.

8. SIPRI, *SIPRI Yearbook 1996, Armaments, Disarmament and International Security* (Oxford: Oxford University Press, 1996).

9. H.C. Engelbrecht and F.C. Hanighen, *Merchants of Death: A Study of the International Armament Industry* (New York: Dodd, Mead, 1934).

10. See Wolpin, *America Insecure* 1991; David Louscher and Michael Salamone, *Technology Transfer and U.S. Security Assistance: The Impact of Licensed Production* (Boulder, CO.: Westview Press, 1987); David Louscher and Michael Salamone, "The Imperative for a New Look at Arms Sales," in *Marketing Security Assistance: New Perspectives on Arms Sales*, ed. David Louscher and Michael Salamone (Lexington, MA: Lexington Books, 1987), 13–40; Paul Hammond, David Louscher, Michael Salamone, and Norman Graham, *The Reluctant Supplier: U.S. Decisionmaking for Arms Sales* (Cambridge, MA: Oelgeschlager, Gunn and Hain, 1983); Peter Mason, *Blood and Iron: Breath of Life or Weapon of Death?* (Victoria, Australia: Penguin Books, 1984); and Stephanie Neuman, *Military Assistance in Recent Wars* (New York: Praeger, 1986).

11. John Keegan's interpretation of Clausewitz's famous dictum. John Keegan, *A History of Warfare* (New York: Vintage Books, 1993), 3.

12. Sherman's definition, of course, was, "War is hell."

13. See especially Brzoska and Pearson, *Arms and Warfare*, 1994; Frederic Pearson, Michael Brzoska, and Christopher Crantz, "The Effect of Arms Transfers on Wars and Peace Negotiations," in SIPRI, *SIPRI Yearbook 1992, Armaments and Disarmament* (Oxford: Oxford University Press, 1992); SIPRI, *The Arms Trade with the Third World* (Stockholm: Almquist and Wiksell, 1971); William Baugh and Michael Squires, "Arms Transfers and the Onset of War Part I: Scalogram Analysis of Transfer Patterns," *International Interactions* 10 (1983): 39–63.

14. See Ian Anthony, Pieter Wezeman, and Siemon Wezeman, "The Trade in Major Conventional Weapons," in SIPRI, *SIPRI Yearbook 1996, Armaments, Disarmament and International Security* (Oxford: Oxford University Press, 1996), 463–536.

15. David Arnold, "Economic Warfare: Targeting Financial Systems as Centers of Gravity," in *Challenge and Response: Anticipating US Military Security Concerns*, ed. Karl Magyar et al. (Maxwell Air Force Base: Air University Press, 1994), 345–62; John Arquilla and David Ronfeldt, eds., *In Athena's Camp: Preparing for Conflict in the Information Age* (Santa Monica, CA: Rand, 1997).

16. Very old studies of the art of war, such as Sun Tzu, dwell at great length on the necessity of utilizing such things as deception, intelligence, and psychological

"operations." See Keegan, *A History of Warfare*, 1993; and Martin van Creveld, *On Future War* (London: Brassey's, 1991).

17. See John Warden, *The Air Campaign: Planning for Combat* (Washington, D.C.: Pergamon-Brassey's, 1989); John Warden, "Employing Air Power in the Twenty-first Century," in *The Future of Air Power in the Aftermath of the Gulf War*, ed. Richard Shultz and Robert Pfaltzgraff (Maxwell Air Force Base: Air University Press, 1992), 57–82; and Richard Reynolds, *Heart of the Storm: The Genesis of the Air Campaign Against Iraq* (Maxwell Air Force Base: Air University Press, 1995).

18. Donald Snow, *Distant Thunder: Patterns of Conflict in the Developing World* (New York: St. Martin's Press, 1993), 57.

19. Adda Bozeman, "War and the Clash of Ideas," *Orbis: A Journal of World Affairs* 20 (1976): 61–102. See also Samuel Huntington, *The Clash of Civilizations and the Remaking of World Order* (New York: Simon and Schuster, 1996); Stephen Blank, Lawrence Grinter, Karl Magyar, Lewis Ware, and Bynum Weathers, *Conflict, Culture, and History: Regional Dimensions* (Maxwell Air Force Base: Air University Press, 1993); Karl Magyar, "Introduction: The Protraction and Prolongation of Wars," in *Prolonged Wars: A Post-Nuclear Challenge*, ed. Karl Magyar and Constantine Danopoulos (Maxwell Air Force Base: Air University Press, 1994).

20. Creveld, *On Future War*, 1991.

21. For example, Michael Zanoni, "The Shopping List: What's Available," in *The International Legal and Illegal Trafficking in Arms*, ed. Peter Unsinger and Harry More (Springfield, IL: Charles C Thomas, 1989); Aaron Karp, "The Arms Trade Revolution: The Major Impact of Small Arms," *Washington Quarterly* (Autumn 1994).

22. It is very difficult to assess the overall global value of this type of weapons trade. If Interarms was taken as an average case, then the value of the 250 listed in Jane's would be about $20 billion per year. Mark Browne, "Conventional Armaments: Mapping Warfare in the Twenty-First Century," in *Global Security Concerns: Anticipating the Twenty-First Century*, ed. Karl Magyar (Maxwell Air Force Base: Air University Press, 1996), 239–58. For recent scholarship on the issue of the "small arms" trade, see Jeffrey Boutwell, Michael Klare, and Laura Reed, eds., *Lethal Commerce: The Global Trade in Small Arms and Light Weapons* (Cambridge, MA: American Academy of Arts and Sciences, 1995); Michael Klare and David Andersen, *A Scourge of Guns: The Diffusion of Small Arms and Light Weapons in Latin America* (Washington, D.C.: Federation of American Scientists, 1996); Jasjit Singh, ed., *Light Weapons and International Security* (New Delhi: Indian Pugwash Society and the British American Security Information Council, 1996); and Project on Light Weapons, *Current Projects on Light Weapons*, Working Paper No. 1 (Washington, D.C.: British American Security Information Council, 1997).

23. Browne, "Conventional Armaments," 1996.

24. Donald Rice, "Air Power in the New Security Environment," in *The Future of Air Power in the Aftermath of the Gulf War*, ed. Shultz and Pfaltzgraff, 1992, 9–16.

25. See Jeffrey Barnett, *Future War: An Assessment of Aerospace Campaigns in 2010* (Maxwell Air Force Base: Air University Press, 1996); Karen Rasler and William Thompson, *The Great Powers and the Global Struggle* (Lexington: University Press of Kentucky, 1994); and Air University, *2025*, 5 vols. (Maxwell Air Force Base: Air University Press, 1996).

26. See Barnett, *Future War*, 1996; Creveld, *On Future War*, 1991; Mario Garza, "Conflict Termination: Every War Must End," in *Challenge and Response: Anticipating US Military Security Concerns*, ed. Karl Magyar et al. (Maxwell Air Force Base: Air University Press, 1996), 413–28; and Gay Hammerman, *Conventional Attrition and Battle Termination Criteria: A Study of War Termination*, DNA-TR-81-224 (Loring, VA: Defense Nuclear Agency, August 1982).

27. A partial exception is Barnett, *Future War* 1996; and Richard Paulsen, *The Role of US Nuclear Weapons in the Post-Cold War Era* (Maxwell Air Force Base: Air University Press, 1994).

28. For a representative discussion of these issues, see Lawrence Freedman, *The Evolution of Nuclear Strategy* (New York: St. Martin's Press); Bernard Brodie, *Strategy in the Missile Age* (Princeton, NJ: Princeton University Press, 1966); Morton Halperin, *Limited War in the Nuclear Age* (New York: John Wiley and Sons, 1963); and Charles Kegley, Jr. and Eugene Wittkopf, *The Nuclear Reader: Strategy, Weapons, War*, 2d ed. (New York: St. Martin's Press, 1989).

29. See Graham Allison, Owen Coté, Richard Falkenrath, and Steven Miller, *Avoiding Nuclear Anarchy* (Cambridge: MIT Press, 1996); David Rosenbaum, "Nuclear Terror," *International Security* 1 (1977): 140–61; Thomas Schelling, "Thinking About Nuclear Terrorism," *International Security* 6 (1982): 61–77; and Paul Leventhal and Yonah Alexander, eds., *Preventing Nuclear Terrorism: The Report and Papers of the International Task Force on Prevention of Nuclear Terrorism* (Lexington, MA: Lexington Books, 1987). See Henry Sokolski, *Fighting Proliferation: New Concerns for the Nineties* (Maxwell Air Force Base: Air University Press, 1996); and Stuart Johnson and William Lewis, *Weapons of Mass Destruction: New Perspectives on Counterproliferation* (Washington, D.C.: National Defense University Press, 1995).

30. See Natalie Goldring, "Bridging the Gap: Light and Major Conventional Weapons in Recent Conflicts" (paper presented at the Annual Meeting of the International Studies Association, Toronto, March 1997), Project on Light Weapons, *Current Projects on Light Weapons*, 1997.

31. See United States Congress, Office of Technology Assessment, *Export Controls and Nonproliferation*, OTA-ISS-596 (Washington, D.C.: U.S. Government Printing Office, 1994); National Research Council, *Proliferation Concerns: Assessing US Efforts to Help Contain Nuclear and other Dangerous Materials and Technologies in the Former Soviet Union* (Washington, D.C.: National Academy Press, 1997); Allison, Coté, Falkenrath, and Miller, *Avoiding Nuclear Anarchy*, 1996; Gary Bertsch and Suzette Grillot, *Arms on the Market* (London: Routledge, 1998).

32. See, especially, Peter Leitner, *Decontrolling Strategic Technology, 1990–1992: Creating the Military Threats of the 21st Century* (New York: University Press of

America, 1995); and R. Inman and Daniel Burton, Jr., "Technology and National Security," in Graham Allison and Gregory Treverton, eds., *Rethinking America's Security: Beyond Cold War to New World Order* (New York: W.W. Norton and Company, 1992).

33. Richard Cupitt and Suzette Grillot, "COCOM is Dead, Long Live COCOM: Persistence and Change in Multilateral Security Institutions," *British Journal of Political Science* 27 (1997): 361–89.

34. Cassady Craft and Suzette Grillot, "Transparency and the Effectiveness of Multilateral Nonproliferation Export Control Regimes: Can Wassenaar Work?" *Southeastern Political Review* (forthcoming); Saferworld, "New Arms Control Regime Risks Being Paper Tiger," *Saferworld Update* (Spring 1996).

35. See Edward Laurance, Siemon Wezeman, and Herbert Wulf, *Arms Watch: SIPRI Report on the First Year of the UN Register of Conventional Arms* (Oxford: Oxford University Press, 1993); Edward Laurance and Tracy Keith, "The United Nations Register of Conventional Arms: On Course in Its Third Year of Reporting," *Nonproliferation Review* 3 (1996): 77–93.

36. Especially when keeping in mind the data bias encountered in Chapter 3.

37. Peter Mason, *Blood and Iron,* 1984, 104.

References

Air University. 1996. *2025.* 5 vols. Maxwell Air Force Base: Air University Press.

Albrecht, Ulrich, Dieter Ernst, Peter Lock, and Herbert Wulf. 1974. Armaments and underdevelopment. *Bulletin of Peace Proposals* 5:173–85.

Allison, Graham, Owen Coté, Richard Falkenrath, and Steven Miller. 1996. *Avoiding nuclear anarchy.* Cambridge, MA: MIT Press.

Altfield, Michael. 1983. Arms races?—and escalation?: A comment on Wallace. *International Studies Quarterly* 27:225–31.

Anderton, Charles. 1992. Toward a mathematical theory of the offensive/defensive balance. *International Studies Quarterly* 36:75–100.

Anthony, Ian, Pieter Wezeman, and Siemon Wezeman. 1996. The trade in major conventional weapons. In Stockholm International Peace Research Institute, *SIPRI Yearbook 1996, Armaments, disarmament and international security.* Oxford: Oxford University Press, 463–536.

Arnold, David. 1994. Economic warfare: Targeting financial systems as centers of gravity. In *Challenge and response: anticipating U.S. military security concerns,* edited by Karl Magyar et al., 345–62. Maxwell Air Force Base: Air University Press.

Arnold, Guy. 1995. *Wars in the Third World since 1945.* London: Cassell.

Arquilla, John, and David Ronfeldt, eds. 1997. *In Athena's camp: Preparing for conflict in the information age.* Santa Monica, CA: Rand.

Avery, William. 1978. Domestic influences on Latin American importation of U.S. armaments. *International Studies Quarterly* 22:121–42.

Barnett, Jeffrey. 1996. *Future war: An assessment of aerospace campaigns in 2010.* Maxwell Air Force Base: Air University Press.

Baugh, William, and Michael Squires. 1983a. Arms transfers and the onset of war part I: Scalogram analysis of transfer patterns. *International Interactions* 10:39–63.

Baugh, William, and Michael Squires. 1983b. Arms transfers and the onset of war part II: Wars in Third World states, 1950–65. *International Interactions* 10:129–41.

Bell, J. Bowyer. 1978. Arms transfers, conflict, and violence at the substate level. In *Arms transfers to the Third World: The military buildup in less industrial countries,* edited by Uri Ra'anan et al., 309–23. Boulder, CO: Westview Press.

Bennett, D. Scott, and Allan Stam. 1996. The duration of interstate wars, 1816–1985. *American Political Science Review* 90:239–57.

Benoit, Emile. 1973. *Defense and economic growth in developing countries.* Lexington, MA: Lexington Books.

Bertsch, Gary, and Suzette Grillot. 1998. *Arms on the market.* London: Routledge.

Bitzinger, Richard. 1994. The globalization of the arms industry: The next proliferation challenge. *International Security* 19:170–98.

Blackaby, Frank, and Thomas Ohlson. 1982. Military expenditure and the arms trade: Problems of data. *Bulletin of Peace Proposals* 13:291–308.

Blainey, Geoffrey. 1988. *The Causes of War.* 3d ed. New York: Free Press.

Blank, Stephen, Lawrence Grinter, Karl Magyar, Lewis Ware, and Bynum Weathers. 1993. *Conflict, culture, and history: Regional dimensions.* Maxwell Air Force Base: Air University Press.

Bloomfield, Lincoln, and Amelia Leiss. 1969. *Controlling small wars; A strategy for the 1970's.* New York: Knopf.

Bobrow, Davis, P. Terrence Hopmann, Roger Benjamin, and Donald Sylvan. 1973. The impact of foreign assistance on national development and international conflict. *Journal of Peace Science* 1:39–60.

Bohrnstedt, George, and David Knoke. 1994. *Statistics for social data analysis.* 3d ed. Itasca, IL: F.E. Peacock.

Boulding, Kenneth. 1963. *Conflict and defense: A general theory.* New York: Harper and Row.

Boutwell, Jeffrey, Michael Klare, and Laura Reed, eds. 1995. *Lethal commerce: The global trade in small arms and light weapons.* Cambridge, MA: American Academy of Arts and Sciences.

Bozeman, Adda. 1976. War and the clash of ideas. *Orbis: A Journal of World Affairs* 20:61–102.

Bracken, Jerome, Moshe Kress, and Richard Rosenthal, eds. 1995. *Warfare modeling.* Danvers, MA: John Wiley and Sons, Inc.

Brodie, Bernard. 1966. *Strategy in the missile age.* Princeton, NJ: Princeton University Press.

Browne, Mark. 1996. Conventional armaments: Mapping warfare in the twenty-first century. In *Global security concerns: Anticipating the twenty-first century,* edited by Karl Magyar, 239–58. Maxwell Air Force Base: Air University Press.

Brzoska, Michael. 1982. Arms transfer data sources. *Journal of Conflict Resolution* 26:77–108.

Brzoska, Michael. 1987. The SIPRI price system. In *SIPRI yearbook 1987: World armaments and disarmament.* Oxford: Oxford University Press.

Brzoska, Michael, and Frederic Pearson. 1994. *Arms and warfare: Escalation, de-escalation, and negotiations.* Columbia: University of South Carolina Press.

Bueno de Mesquita, Bruce. 1981. *The war trap.* New Haven, CT: Yale University Press.

Bueno de Mesquita, Bruce, and David Lalman. 1992. *War and reason.* New Haven, CT: Yale University Press.

Burns, Arthur. 1959. A graphical approach to some problems of the arms race, *Journal of Conflict Resolution* 3:326–42.

Callahan, Leslie, Jr. 1983. The Need for a Multidisciplinary Modeling Language in Military Science and Engineering. In *Modeling and simulation of land combat,* edited by Leslie Callahan, Jr. Atlanta: Georgia Tech Research Institute.

Carnegie Commission on Preventing Deadly Conflict. 1997. *Preventing deadly conflict: Final report.* New York: Carnegie Corporation of New York, December.

Carr, E.H. 1946. *The twenty years crisis, 1919–1939: An introduction to the study of international relations.* London: Macmillan.

Carter, Barry. 1988. *International economic sanctions: Improving the haphazard U.S. legal regime.* New York: Cambridge University Press.

Catrina, Christian. 1988. *Arms transfers and dependence.* New York: United Nations for Disarmament Research.

Catrina, Christian. 1994. Main directions of research in the arms trade. *Annals of the American Academy of Political and Social Science* 535:190–205.

Chuyev, Yu. 1968. *Fundamentals of operations research in combat materiel and weaponry,* Vol. 2. Wright Patterson AFB, OH: Foreign Technology Division.

Chuyev, Yu. and Yu. Mikhaylov. 1980. *Forecasting in military affairs: A Soviet view.* Washington, D.C.: U.S. Government Printing Office.

Cioffi-Revilla, Claude. 1991. On the likely magnitude, extent, and duration of an Iraq-U.S. war. *Journal of Conflict Resolution* 35:387–411.

Clausewitz, Carl von. 1832–35. *Vom Krieg* [On War]. Translated by Michael Howard and P. Paret. Princeton, NJ: Princeton University Press, 1976.

Cox, David. 1989. *Analysis of binary data.* 2d ed. New York: Chapman and Hall.

Craft, Cassady, and Suzette Grillot. Forthcoming. Transparency and the effectiveness of multilateral nonproliferation export control regimes: Can Wassenaar work? *Southeastern Political Review.*

Creveld, Martin van. 1991. *On future war.* London: Brassey's.

Cupitt, Richard, and Suzette Grillot. 1997. COCOM is dead, long live COCOM: Persistence and change in multilateral security institutions. *British Journal of Political Science* 27:361–89.

Davis, Paul. 1992. Variable-resolution combat modeling. In *Proceedings of the 1991 Callaway workshop: Modeling, simulation, and gaming for restructuring the U.S. armed forces,* edited by Leslie Callahan, Jr., and Ross Gagliano. Pine Mountain, GA: U.S. Army Missile Command.

Davis, Paul. 1995. An introduction to variable resolution modeling. In *Warfare Modeling,* edited by Jerome Bracken, Moshe Kress, and Richard Rosenthal, 5–37. Washington, DC: Military Operations Research Society (MORS).

Diehl, Paul. 1983. Arms races and escalation: A closer look, *Journal of Peace Research* 20:205–12.

Diehl, Paul. 1985. Arms races to war: Testing some empirical linkages. *Sociological Quarterly* 96:331–49.

Diehl, Paul, and Jean Kingston. 1987. Messenger or message?: Military buildups and the initiation of conflict. *Journal of Politics* 49:801–13.

Dolian, James. 1973. Military coups and the allocation of national resources: An examination of 34 Sub-Saharan African Nations. Unpublished Ph.D. dissertation, Northwestern University, Evanston, IL.

Dongen, Martin van, and Joost Kos. 1995. The analysis of ship air defense: The simulation model SEAROADS. In *Warfare Modeling,* edited by Jerome Bracken et al., 145–63. Washington, D.C.: John Wiley and Sons.

Doxey, Margaret. 1987. *International sanctions in contemporary perspective.* New York: St. Martin's Press.

Dressler, David. 1991. Beyond correlations: Toward a causal theory of war. *International Studies Quarterly* 35:337–55.

Dunnigan, James. 1988. *How to make war.* New York: William Morrow.

Dupuy, Ernest, and Trevor Dupuy. 1986. *The encyclopedia of military history from 3500 BC to the present.* New York: Harper and Row.

Dupuy, Trevor. 1979. *Numbers, predictions and war: Using history to evaluate combat factors and predict the outcome of battles.* Indianapolis: Bobbs-Merrill.

Dupuy, Trevor. 1983. *Analysis of factors that have influenced outcomes of battles and wars: A data base of battles and engagements, final report.* 6 vols. Dunn Loring, VA: Historical Evaluation and Research Organization.

Duvall, Raymond. 1976. An appraisal of the methodological and statistical procedures of the correlates of war project. In *Quantitative international politics: An appraisal,* edited by Francis Hoole and Dina Zinnes, 67–98. New York: Praeger.

Eckhardt, William, and Edward Azar. 1978. Major world conflicts and interventions, 1945–1975. *International Interactions* 5:75–109.

Einaudi, Luigi, Hans Heymann, Jr., David Ronfeldt, and Cesar Sereseres. 1973. *Arms transfers to Latin America: Toward a policy of mutual respect.* Report R-1173-DOS. Santa Barbara, CA: Rand.

Engelbrecht, H.C., and F.C. Hanighen. 1934. *Merchants of death: A study of the international armament industry.* New York: Dodd, Mead.

Engle, J.H. 1954. A verification of Lanchester's Law. *Operations Research* 2.

Epstein, Joshua. 1984. *Measuring military power: The soviet air threat to Europe.* Princeton, NJ: Princeton University Press.

Epstein, Joshua. 1985. *The calculus of conventional war: Dynamic analysis without Lanchester theory.* Washington, D.C.: Brookings Institution.

Epstein, Joshua. 1987. *Strategy and force planning: The case of the Persian Gulf.* Washington, D.C.: Brookings Institution.

Epstein, Joshua. 1989. The 3:1 rule, the adaptive dynamic model, and the future of security studies. *International Security* 13:90–127.

Epstein, Joshua. 1990. *Conventional force reductions: A dynamic assessment.* Washington, D.C.: Brookings Institution.

Epstein, Joshua, Kim Holmes, John Mearsheimer, and Barry Posen. 1988. The European conventional balance. *International Security* 12:152–202.

Fei, Edward. 1979. Understanding arms transfers and military expenditures: Data problems. In *Arms transfers in the modern world*, edited by Stephanie Neuman and Robert Harkavy, 37–48. New York: Praeger.

Foster, James. 1978. New conventional weapons technologies: Implications for the Third World. In *Arms transfers to the Third World: The military buildup in less industrial countries*, edited by Uri Ra'anan, Robert Pfaltzgraff, and Geoffrey Kemp, 65–84, Boulder, CO: Westview Press.

Fowler, Bruce. 1992. A *Perestroika* program for modeling and simulation tools and techniques. In *Proceedings of the 1991 Callaway workshop: Modeling, simulation, and gaming for restructuring the U.S. armed forces*, edited by Leslie Callahan, Jr., and Ross Gagliano. Pine Mountain, GA: U.S. Army Missile Command.

Frederikson, Peter, and Robert Looney. 1982. Defense expenditures and economic growth in developing countries: Some further empirical evidence. *Journal of Economic Development*, 7:113–26.

Freedman, Lawrence. *The evolution of nuclear strategy*. New York: St. Martin's Press.

Garza, Mario. 1996. Conflict termination: Every war must end. In *Challenge and response: Anticipating US military security concerns*, edited by Karl Magyar et al., 413–28. Maxwell Air Force Base: Air University Press.

Gavril, Fanica. 1995. A synthesis of methods for simulating vulnerability of armored vehicles. In *Warfare Modeling*, edited by Jerome Bracken, Moshe Kress, and Richard Rosenthal, 101–20. Washington, D.C.: Military Operations Research Society (MORS).

Gerner, Debbie. 1982. A statistical study of arms transfers and domestic conflict in 57 African and West Asian nations, 1963–1978. Unpublished Ph.D. dissertation, Northwestern University, Evanston, IL.

Gerner, Debbie. 1983. Arms transfers to the third world: Research on patterns, causes and effects." *International Interactions* 10:5–37.

Gilpin, Robert. 1989. The theory of hegemonic war. In *The origin and prevention of major wars*, edited by R. Rotberg and Theodore Rabb. Cambridge: Cambridge University Press.

Gochman, Charles, and Zeev Maoz. 1984. Militarized interstate disputes, 1816–1976: Procedures, patterns, and insights. *Journal of Conflict Resolution* 28:585–615.

Goldring, Natalie. 1997. Bridging the gap: Light and major conventional weapons in recent conflicts. Paper presented at the Annual Meeting of the International Studies Association, Toronto, March.

Greene, William. 1997. *Econometric analysis*. 3d ed. Upper Saddle River, NJ: Prentice-Hall.

Gurr, Ted. 1971. *Why men rebel*. Princeton: Princeton University Press.

Halperin, Morton. 1963. *Limited war in the nuclear age*. New York: John Wiley and Sons.

Hammerman, Gay. 1982. *Conventional attrition and battle termination criteria: A study of war termination*. DNA-TR-81-224. Loring, VA: Defense Nuclear Agency, August.

Hammond, Kenneth, and James Householder. 1967. *Introduction to the statistical method: Foundations and use in the behavioral sciences*. 3d ed. New York: Knopf.

Hammond, Paul, David Louscher, Michael Salamone, and Norman Graham. 1983. *The reluctant supplier: U.S. decisionmaking for arms sales*. Cambridge, MA: Oelgeschlager, Gunn and Hain.

Hanushek, Eric, and John Jackson. 1977. *Statistical methods for social scientists*. New York: Academic Press.

Harkavy, Robert. 1975. *The arms trade and international systems*. Cambridge, MA: Ballinger Publishing Company.

Harkavy, Robert. 1984. Recent wars in the arc of crisis: Lessons for defense planners. In *Defense planning in less-industrialized states: The Middle East and South Asia*, edited by Stephanie Neuman, 275–300. Lexington, MA: D.C. Heath.

Harkavy, Robert, and Stephanie Neuman, eds. 1985a. *The lessons of recent wars in the Third World, volume I: Approaches and case studies*. Lexington, MA: Lexington.

Harkavy, Robert, and Stephanie Neuman, eds. 1985b. *The lessons of recent wars in the Third World, volume II: Comparative dimensions*. Lexington, MA: Lexington.

Hartley, Dean. 1992. Mathematical modeling of historical combat data. In *Proceedings of the 1991 Callaway workshop: Modeling, simulation, and gaming for restructuring the U.S. armed forces*, edited Leslie Callahan, Jr., and Ross Gagliano. Pine Mountain, GA: U.S. Army Missile Command.

Hartung, William. 1994. *And weapons for all: How America's multibillion-dollar arms trade warps our foreign policy and subverts democracy at home.* New York: HarperCollins.

Hartung, William. 1997. Weapons at war: Patterns of arms deliveries to recent conflicts. Presented at the annual meeting of the International Studies Association, March 18–22.

Helmbold, Robert. 1964. Some observations on the use of Lanchester's theory for prediction. *Operations Research* 12:778–81.

Helmbold, Robert, and Aqeel Khan. 1986. *Combat history analysis study effort (CHASE) progress report for the period August 1984–June 1985.* Washington, D.C.: Requirements and Resources Directorate, U.S. Army Concepts and Analysis Agency, August.

Hibbs, Douglas. 1973. *Mass political violence: A cross-national causal analysis.* New York: Wiley-Interscience.

Hillenbrand, Martin. 1949. *Power and morals.* New York: Columbia University Press.

Hillestad, Richard, and Mario Juncosa. 1995. Cutting some trees to see the forest: On aggregation and disaggregation in combat models. In *Warfare modeling*, edited by Jerome Bracken, Moshe Kress, and Richard Rosenthal, 37–62. Washington, D.C.: Military Operations Research Society (MORS).

Hopf, Ted. 1994. *Peripheral visions: Deterrence theory and American foreign policy in the Third World, 1965–1990.* Ann Arbor: University of Michigan Press.

Houweling, Henk, and Jan Siccama. 1988. *Studies of war.* Dordrecht, The Netherlands: Martinus Nijhoff.

Howard, Michael. 1976. *War in European history.* London: Oxford University Press.

Hufbauer, Gary, and Jeffrey Schott. 1985. *Economic sanctions reconsidered: History and current policy.* Washington, D.C.: Institute for International Economics.

Hughes, Wayne. 1995. A Salvo model of warships in missile combat used to evaluate their staying power. In *Warfare Modeling*, edited by Jerome Bracken, Moshe Kress, and Richard Rosenthal, 121–44. Washington, D.C.: Military Operations Research Society (MORS).

Huntington, Samuel. 1969. *Political order in changing societies.* New Haven, CT: Yale University Press.

Huntington, Samuel. 1996. *The clash of civilizations and the remaking of world order.* New York: Simon and Schuster.

Huth, Paul, and Bruce Russett. 1984. What makes deterrence work?: Cases from 1900–1980. *World Politics* 36:496–526.

Huth, Paul, and Bruce Russett. 1988. Deterrence failure and crisis escalation. *International Studies Quarterly* 32:29–45.

Huth, Paul, and Bruce Russett. 1990. Testing deterrence theory: Rigor makes a difference. *World Politics* 42:466–501.

Huth, Paul, and Bruce Russett. 1993. General deterrence between enduring rivals: Testing three competing models. *American Political Science Review* 87:61–73.

Huth, Paul, D. Scott Bennett, and Christopher Gelpi. 1992. System uncertainty, risk propensity, and international conflict among the great powers. *Journal of Conflict Resolution* 36:478–517.

Inman, R., and Daniel Burton, Jr. 1992. Technology and national security. In *Rethinking America's security: Beyond cold war to new world order*, edited by Graham Allison and Gregory Treverton. New York: W.W. Norton and Company.

Intriligator, Michael, and Dagobert Brito. 1976a. Strategy, arms races, and arms control. In *Mathematical systems in international relations*, edited by J. Gillespie and Dina Zinnes. New York: Praeger.

Intriligator, Michael, and Dagobert Brito. 1976b. Formal models of arms races. *Journal of Peace Science* 2:77–88.

Intriligator, Michael, and Dagobert Brito. 1984. Can arms races lead to the outbreak of war? *Journal of Conflict Resolution* 28:63–84.

Intriligator, Michael, and Dagobert Brito. 1986. Arms Races and Instability. *Journal of Strategic Studies* 9:113–31.

Intriligator, Michael, and Dagobert Brito. 1989. Richardsonian arms race models. In *Handbook of war studies*, edited by Manus Midlarsky. Boston: Unwin Hyman.

Isaard, Walter, and Charles Anderton. 1985. Arms race models: A survey and synthesis. *Conflict Management and Peace Science* 8:27–98.

Jervis, Robert. 1978. Cooperation under the security dilemma. *World Politics* 30:186–214.

Johnson, Stuart, and William Lewis. 1995. *Weapons of mass destruction: New perspectives on counterproliferation.* Washington, D.C.: National Defense University Press.

Kaplan, Stephen. 1975. U.S. arms transfers to Latin America, 1945–1974. *International Studies Quarterly* 19:399–431.

Karp, Aaron. 1994. The arms trade revolution: The major impact of small arms. *Washington Quarterly* (Autumn).

Kaufmann, William. 1983. The arithmetic of force planning. In *Alliance security: NATO and the no-first-use question*, edited by John Steinbruner and Leon Sigal. Washington, D.C.: Brookings Institution.

Keegan, John. 1993. *A history of warfare.* New York: Vintage Books.

Kegley, Charles, Jr., and Eugene Wittkopf. 1989. *The nuclear reader: Strategy, weapons, war.* 2d ed. New York: St. Martin's Press.

Kemp, Geoffrey, with Stephen Miller. 1979. Arms transfers phenomenon. In *Arms transfers and American foreign policy*, edited by Andrew Pierre. New York: New York University Press.

Kende, Istvan. 1971. Twenty-five years of local wars. *Journal of Peace Research* 8:5–27.

Kende, Istvan. 1978. Wars of ten years (1967–1976). *Journal of Peace Research* 15:227–41.

King, Gary. 1989. *Unifying political methodology.* Cambridge: Cambridge University Press.

Kinsella, David, 1994. Conflict in context: Arms transfers and third world rivalries during the Cold War. *American Journal of Political Science.* 38:557–81.

Kinsella, David, and Herbert Tillema. 1995. Arms and aggression in the Middle East. *Journal of Conflict Resolution* 39:306–29.

Klare, Michael, and David Andersen. 1996. *A scourge of guns: The diffusion of small arms and light weapons in Latin America.* Washington, D.C.: Federation of American Scientists.

Kolodziej, Edward. 1979. Measuring French arms transfers. *Journal of Conflict Resolution* 23:195–227.

Krause, Keith. 1991. Military statecraft: Power and influence in Soviet and American arms transfer relationships. *International Studies Quarterly* 35:313–36.

Lambelet, John. 1975. A numerical model of the Anglo-German dreadnought race. *Peace Science Society (International), Papers* 24:29–48.

Lanchester, Frederick. 1916. *Aircraft in warfare: The dawn of the fourth arm.* London: Constable.

Laurance, Edward. 1992. *The international arms trade.* New York: Lexington Books.

Laurance, Edward. 1993. *Arms watch: SIPRI report on the first year of the UN register of conventional arms.* Oxford: Oxford University Press.

Laurance, Edward, and Tracy Keith. 1996. The United Nations register of conventional arms: On course in its third year of reporting. *Nonproliferation Review* 3:77–93.

Laurance, Edward, and Joyce Mullen. 1987. Assessing and analyzing international arms trade data. In *Marketing security assistance: New perspectives on arms sales*, edited by David Louscher and Michael Salamone, 79–98. Lexington, MA: Lexington Books.

Laurance, Edward, and Ronald Sherwin. 1978. Understanding arms transfers through data analysis. In *Arms Transfers to the Third World: The Military Buildup in Less Industrial Countries*, edited by Uri Ra'anan, Robert Pfaltzgraff, and Geoffrey Kemp, 87–105. Boulder, CO: Westview Press.

Laurance, Edward, Siemon Wezeman, and Herbert Wulf. 1993. *Arms watch: SIPRI report on the first year of the UN register of conventional arms.* Oxford: Oxford University Press.

Leiss, Amelia, with Geoffrey Kemp et al. 1970. *Arms and local conflict.* Vol. 3 of *Arms transfers to less developed countries.* Cambridge: MIT Center for International Studies.

Leitner, Peter. 1995. *Decontrolling strategic technology, 1990–1992: Creating the military threats of the 21st century.* New York: University Press of America.

Lepingwell, John. 1987. The laws of combat?: Lanchester reexamined. *International Security* 12:1, 89–139.

Leventhal, Paul, and Yonah Alexander, eds. 1987. *Preventing nuclear terrorism: The report and papers of the international task force on prevention of nuclear terrorism.* Lexington, MA: Lexington Books.

Louscher, David, and Michael Salamone. 1987a. *Technology transfer and U.S. security assistance: The impact of licensed production.* Boulder, CO: Westview Press.

Louscher, David, and Michael Salamone. 1987b. The imperative for a new look at arms sales. In *Marketing Security Assistance: New Perspectives on Arms Sales,* edited by David Louscher and Michael Salamone, 13–40. Lexington, MA: Lexington Books.

Magyar, Karl. 1994. Introduction: The protraction and prolongation of wars. In *Prolonged wars: A post-nuclear challenge,* edited by Karl Magyar and Constantine Danopoulos. Maxwell Air Force Base: Air University Press.

Majeski, Steven, and David Jones. 1981. Arms race modeling: Causality analysis and model specification. *Journal of Conflict Resolution* 25:259–88.

Mako, William. 1983. *U.S. ground forces and the defense of central Europe.* Washington, D.C.: Brookings Institution.

Mandel, Robert. 1997. Exploding myths about global arms transfers. Presented at the annual meeting of the International Studies Association, March 18–22.

Maniruzzaman, Talukder. 1992. Arms transfers, military coups, and military rule in developing states. *Journal of Conflict Resolution* 36:733–55.

Maoz, Zeev. 1982. *Paths to conflict.* Boulder, CO: Westview Press.

Maoz, Zeev. 1989. Power, capabilities, and paradoxical conflict outcomes. *World Politics* 41:239–66.

Mason, Peter. 1984. *Blood and iron: Breath of life or weapon of death?* Victoria, Australia: Penguin Books.

McQuie, Robert. 1988a. *Historical characteristics of combat for wargames (Benchmarks).* CAA-RP-87-2. Washington, D.C.: U.S. Army Concepts and Analysis Agency, July.

McQuie, Robert. 1988b. *A set of templates for evaluating wargames (Benchmarks).* Washington, D.C.: U.S. Army Concepts and Analysis Agency, October.

Mearsheimer, John. 1989. Assessing the conventional balance: The 3:1 rule and its critics. *International Security* 13:54–89.

Mearsheimer, John, Barry Posen, Eliot Cohen, Steven Zaloga, Malcolm Chalmers, and Lutz Unterseher. 1989. Correspondence, *Inter-national Security* 13:128–79.

Meier, Kenneth, and Jeffrey Brudney. 1987. *Applied statistics for public administration.* Rev. ed. Pacific Grove, CA: Brooks/Cole Publishing Company.

Merritt, Richard, and Dina Zinnes. 1988. Validity of power indices. *International Interactions* 14:141–51.

Milstein, Jeffrey. 1972. American and Soviet influence, balance of power, and Arab-Israeli violence. In *Peace, War, and Numbers,* edited by Bruce Russett. Beverly Hills: Sage.

Mintz, Alex. 1986a. Arms exports as an action-reaction process. *The Jerusalem Journal of International Relations* 8:102–13.

Mintz, Alex. 1986b. Arms imports as an action-reaction process: An empirical test of six pairs of developing nations. *International Interactions* 12:229–43.

Modelski, George, and William Thompson. 1988. *Seapower in global politics, 1494–1993.* Seattle: University of Washington Press.

Montgomery, Douglas, and Elizabeth Peck. 1982. *Introduction to linear regression analysis.* New York: Wiley.

Morgenthau, Hans. 1948. *Politics among nations: The struggle for power and peace.* New York: Knopf.

Morrow, James. 1989. A twist of truth: A reexamination of the effects of arms races on the occurrence of war. *Journal of Conflict Resolution* 33:500–29.

Moul, William. 1989. Measuring the "balances of power": A look at some numbers. *Review of International Studies* 15:101–21.

Moul, William. 1994. Predicting the severity of great power war from its extent. *Journal of Conflict Resolution* 38:160–69.

Mueller, John. 1991. Changing attitudes to war: The impact of the First World War. *British Journal of Political Science* 21:25–50.

National Research Council. 1997. *Proliferation concerns: Assessing U.S. efforts to help contain nuclear and other dangerous materials and technologies in the former Soviet Union.* Washington, D.C.: National Academy Press.

Neuman, Stephanie. 1979. Arms transfers and economic development: Some research and policy issues. In *Arms Transfers in the Modern World*, edited by Stephanie Neuman and Robert E. Harkavy. New York: Praeger.

Neuman, Stephanie. 1984. International stratification and Third World military industries. *International Organization* 38:167–97.

Neuman, Stephanie. 1986a. The arms trade in recent wars. *Journal of International Affairs* 40:77–99.

Neuman, Stephanie. 1986b. *Military assistance in recent wars.* New York: Praeger.

Neuman, Stephanie. 1986c. The role of military assistance in recent wars. In *The lessons of recent wars in the Third World, volume II*, edited by Stephanie Neuman and Robert Harkavy, 115–56. Lexington, MA: D.C. Heath.

Neuman, Stephanie. 1987a. Third World military industries: Capabilities and constraints in recent wars. In *The lessons of recent wars in the Third World, volume II*, edited by Robert Harkavy and Stephanie Neuman. Lexington, MA: Lexington Books.

Neuman, Stephanie. 1987b. The role of military assistance in recent wars. In *The lessons of recent wars in the Third World, volume II*, edited by Robert Harkavy and Stephanie Neuman. Lexington, MA: Lexington Books.

Neuman, Stephanie. 1988. Arms, aid and the superpowers. *Foreign Affairs* 66:1044–66.

Oberg, Jan. 1975. Third World armament: Domestic arms production in Israel, South Africa, Brazil, Argentina and India. *Instant Research on Peace and Violence* 5:1050–75.

Ohlson, Thomas. 1982. Third World arms exporters—A new facet of the global arms race. *Bulletin of Peace Proposals* 13:201–20.

O'Neal, John. 1989. Measuring the material base of the contemporary East-West balance of power. *International Interactions* 15:177–96.

Organski, A.F.K., and Jacek Kugler. 1980. *The war ledger.* Chicago: University of Chicago Press.

Osipov, Mikhail. 1915. *Vliyaniye chislennosti srazhayushchikhsya storon na ikh poteri* [The influence of the numerical strength of engaged sides on their casualties]. Translated by Robert Helmbold and Allan Rehm in *Warfare modeling*, edited by Bracken et al., 289–344. Davers, MA: John Wiley and Sons.

Paul, T.V. 1994. *Asymmetric conflicts: War initiation by weaker powers.* New York: Cambridge University Press.

Paulsen, Richard. 1994. *The role of U.S. nuclear weapons in the post–Cold War era.* Maxwell Air Force Base: Air University Press.

Pearson, Frederic. 1981. An analysis of the linkage between arms transfers and subsequent military intervention. Center for International Studies, Occasional Paper 8102.

Pearson, Frederic, Michael Brzoska, and Christopher Crantz. 1992. The effect of arms transfers on wars and peace negotiations. In Stockholm International Peace Research Institute, *SIPRI Yearbook 1992, armaments and disarmament.* Oxford: Oxford University Press.

Pierre, Andrew, ed. 1979. *Arms transfers and American foreign policy.* New York: New York University Press.

Posen, Barry. 1984–5. Measuring the European conventional balance: Coping with complexity in threat assessment. *International Security* 9:56–81.

Powell, Robert. 1996. Stability and the distribution of power. *World Politics* 48:239–67.

Project on Light Weapons. 1997. *Current projects on light weapons.* Working Paper No. 1. Washington, D.C.: British American Security Information Council.

Quandt, William. 1978. Influence through arms supply: The U.S. experience in the Middle East. In *Arms transfers to the Third World: The military buildup in less industrial countries*, edited by Uri Ra'anan, Robert Pfaltzgraff, and Geoffrey Kemp, 121–30. Boulder, CO: Westview Press.

Quester, George. 1987. Six causes of war. In *The Future of Nuclear Deterrence* edited by George Quester. Lexington, MA: Lexington Books.

Ra'anan, Uri. 1978. Soviet arms transfers and the problem of political leverage. In *Arms transfers to the Third World: The military buildup in less industrial countries*, edited by Uri Ra'anan, Robert Pfaltzgraff, and Geoffrey Kemp, 131–56. Boulder, CO: Westview Press.

Ra'anan, Uri, Robert Pfaltzgraff, and Geoffrey Kemp, eds. 1978. *Arms transfers to the Third World: The military buildup in less industrial countries*. Boulder, CO: Westview Press.

Rapoport, Anatoli. 1957. Lewis F. Richardson's mathematical theory of war. *Journal of Conflict Resolution* 1:249–304.

Rapoport, Anatoli. 1960. *Fights, games and debates*. Ann Arbor: University of Michigan Press.

Rasler, Karen, and William Thompson. 1994. *The great powers and the global struggle*. Lexington: University Press of Kentucky.

Rattinger, Hans. 1976. From war to war to war: Arms races in the Middle East. *International Studies Quarterly* 20:501–31.

Raymond, A.D. 1991. Assessing combat power: A methodology for tactical battle staffs. Fort Leavenworth, KS: Army Command and General Staff College.

Reynolds, Richard. 1995. *Heart of the storm: The genesis of the air campaign against Iraq*. Maxwell Air Force Base: Air University Press.

Rice, Donald. 1992. Air power in the new security environment. In *The future of air power in the aftermath of the Gulf War*, edited by Richard Shultz and Robert Pfaltzgraff, 9–16. Maxwell Air Force Base: Air University Press.

Richardson, Lewis. 1939. Generalized foreign politics. *British Journal of Psychology Monographs Supplement* 23.

Richardson, Lewis. 1951. Could an arms race end without fighting? *Nature* 4274:567–69.

Richardson, Lewis. 1960a. *Arms and insecurity*. Pittsburgh: Boxwood.

Richardson, Lewis. 1960b. *Statistics of deadly quarrels*. Chicago: University of Chicago Press.

Ropp, Theodore. 1964. *War in the modern world*. New York: Collier Books.

Rosen, Steven. 1979. The proliferation of new land-based technologies: Implications for local military balances. In *Arms transfers in the modern world*, edited by Stephanie Neuman and Robert E. Harkavy, 109–30. New York: Praeger.

Rosenbaum, David. 1977. Nuclear terror. *International Security* 1:140–61.

Rummel, R.J. 1975. *Understanding conflict and war*. Beverly Hills: Sage.

Saferworld. 1996. New arms control regime risks being paper tiger. *Saferworld Update*. Spring.

Sanjian, Gregory. 1987. *Arms transfers to the Third World: Probability models of superpower decisionmaking*. Boulder, CO: Lynne Rienner.

Sanjian, Gregory. 1988a. Arms export decision-making: A fuzzy control model. *International Interaction* 14:243–65.

Sanjian, Gregory. 1988b. Fuzzy set theory and U.S. arms transfers: Modeling the decision-making process. *American Journal of Political Science* 32:1018–46.

Sanjian, Gregory. 1991. Great power arms transfers: Modeling the decision-making processes of hegemonic, industrial, and restrictive exporters. *International Studies Quarterly* 35:173–93.

Sanjian, Gregory. 1997. A model of an arms trade system. Presented at the annual meeting of the International Studies Association, March 18–22.

Schelling, Thomas. 1982. Thinking about nuclear terrorism. *International Security* 6:61–77.

Schmitter, Philippe. 1973. Foreign military assistance, national military spending and military role in Latin America. In *Military Role in Latin America*, edited by Philippe Schmitter. Beverly Hills: Sage.

Schoultz, Lars. 1988. *National security and United States policy toward Latin America*. Princeton, NJ: Princeton University Press.

Schrodt, Philip. 1983. Arms transfers and international behavior in the Arabian Sea area. *International Interactions* 10:101–27.

Schwarz, Anne Naylor. 1987. Arms transfers and the development of second-level arms industries. In *Marketing Security Assistance: New Perspectives on Arms Sales*, edited by David Louscher and Michael Salamone, 101–30. Lexington, MA: Lexington Books.

Seldes, George. 1934. *Iron, blood and profits: An exposure of the world-wide munitions racket*. New York: Harper and Brothers.

Shephard, R., D. Hartley, P. Haysman, and L. Thorpe. 1988. *Applied operations research: Examples from defense assessment*. New York: Plenum Press.

Sherwin, Ronald. 1983. Controlling instability and conflict through arms transfers: Testing a policy assumption. *International Interactions* 10:65–99.

Sherwin, Ronald, and Edward Laurance. 1979. Arms transfers and military capability: Measuring and evaluating conventional arms transfers. *International Studies Quarterly* 23:360–89.

Shoemaker, Christopher, and John Spanier. 1984. *Patron-client state relationships: Multilateral crises in the nuclear age*. New York: Praeger.

Singer, J. David. 1958. Threat-perception and the armament-tension dilemma. *Journal of Conflict Resolution* 2:90–105.

Singer, J. David. 1962. *Deterrence, arms control, and disarmament*. Columbus: Ohio State University Press.

Singer, J. David, ed. 1979a. *Correlates of war I: Research origins and rationale*. New York: Free Press.

Singer, J. David, ed. 1979b. *Correlates of war II: Testing some realpolitik models*. New York: Free Press.

Singer, J. David, ed. 1979c. *Explaining war: Selected papers from the correlates of war project*. Beverly Hills: Sage.

Singer, J. David, and Melvin Small, eds. 1972. *The wages of war: 1816–1965, a statistical handbook*. New York: John Wiley and Sons.

Singer, J. David, and Melvin Small. 1993. Correlates of war project: international and civil war data, 1816–1992 [Computer file]. Ann Arbor, MI: J. David Singer and Melvin Small [producers]. Ann Arbor, MI: Inter-university Consortium for Political and Social Research [distributor], 1994.

Singer, J. David, and Paul Diehl, eds. 1990. *Measuring the correlates of war*. Ann Arbor: University of Michigan Press.

Singh, Jasjit, ed. 1996. *Light weapons and international security*. New Delhi: Indian Pugwash Society and the British American Security Information Council.

Siverson, Randolph, and Paul Diehl. 1989. Arms races, the conflict spiral, and the onset of war. In *Handbook of War Studies*, edited by Manus Midlarsky. Boston: Unwin Hyman.

Sköns, E. 1992. Sources and methods. In *SIPRI yearbook 1992: World armaments and disarmament*. Oxford: Oxford University Press.

Small, Melvin, and J. David Singer. 1982. *Resort to arms: International and civil wars, 1816–1980*. Beverly Hills: Sage.

Snider, Lewis. 1979. Arms transfers and recipient cooperation with supplier policy preferences. *International Interaction* 5:241–66.

Snider, Lewis. 1987. Do arms exports contribute to savings in defense spending? A cross-sectional pooled time series analysis. In *Marketing security assistance: New perspectives on arms sales*, edited by David Louscher and Michael Salamone, 41–64. Lexington, MA: Lexington Books.

Snow, Donald. 1993. *Distant thunder: Patterns of conflict in the developing world*. New York: St. Martin's Press.

Snow, Donald. 1997. *Distant thunder: Patterns of conflict in the developing world*. 2d ed. Armonk, NY: M.E. Sharpe.

Sokolski, Henry. 1996. *Fighting proliferation: New concerns for the nineties*. Maxwell Air Force Base: Air University Press.

Stam, Allan. 1996. *Win, lose, or draw: Domestic politics and the crucible of war*. Ann Arbor: University of Michigan Press.

Stanley, John, and Maurice Pearton. 1972. *The international trade in arms*. New York: Praeger.

Starr, Harvey. 1994. Revolution and war: Rethinking the linkage between internal and external conflict. *Political Research Quarterly* 47:481–507.

Starr, Harvey, and Benjamin Most. 1980. Diffusion, reinforcement, geopolitics, and the spread of war. *American Political Science Review* 74:609–36.

Staudenmaier, William. 1985. Iran-Iraq (1980–). In *The lessons of recent wars in the Third World, volume I*, edited by Robert Harkavy and Stephanie Neuman, 211–38. Lexington, MA: D.C. Heath.

Stockholm International Peace Research Institute. 1971. *The arms trade with the Third World.* Stockholm: Almquist and Wiksell.

Stockholm International Peace Research Institute. 1996. *SIPRI yearbook 1996, armaments, disarmament and international security.* Oxford: Oxford University Press.

Stockholm International Peace Research Institute. 1997. SIPRI Conventional weapons database, 1950–1997 [Computer file]. Stockholm: Stockholm International Peace Research Institute [producers].

Stoll, Richard, and Michael Ward. 1989. *Power in world politics.* Boulder, CO: Lynne Rienner.

Sylvan, Donald. 1976. Consequences of sharp military assistance increases for international conflict and cooperation. *Journal of Conflict Resolution* 20:609–36.

Taylor, James. 1981. *Force-on-force attrition modeling.* Arlington, VA: Operations Research Society of America.

Taylor, James. 1983. *Lanchester models of warfare.* 2 vols. Arlington, VA: Operations Research Society of America.

Thompson, William. 1993. The consequences of war. *International Interactions* 19:125–47.

U.S. Congressional Budget Office. 1988. *U.S. ground forces and the conventional balance in Europe.* Washington, D.C.: U.S. Government Printing Office.

U.S. General Accounting Office (GAO). 1980. *Models, data, and war: A critique of the foundation for defense analysis,* PAD-80-21, Washington, D.C.: U.S. Government Printing Office.

United States Congress, Office of Technology Assessment. 1994. *Export controls and nonproliferation.* OTA-ISS-596. Washington, D.C.: U.S. Government Printing Office.

Wallace, Michael. 1973. *War and rank among nations.* Lexington, MA: D.C. Heath.

Wallace, Michael. 1982a. Arms races and escalation. *International Studies Quarterly* 26:37–56.

Wallace, Michael. 1982b. Arms races and escalation: Some new evidence. *Journal of Conflict Resolution* 23:3–16.

Waltz, Kenneth. 1959. *Man, the state, and war: A theoretical analysis.* New York: Columbia University Press.

Waltz, Kenneth. 1979. *Theory of international politics.* Reading, MA: Addison-Wesley.

Warden, John. 1989. *The air campaign: Planning for combat.* Washington, D.C.: Pergamon-Brassey's.

Warden, John. 1992. Employing air power in the twenty-first century. In *The Future of Air Power in the Aftermath of the Gulf War,* edited by Richard Shultz and Robert Pfaltzgraff, 57–82. Maxwell Air Force Base: Air University Press.

Weede, Eric. 1980. Arms races and escalation: Some persisting doubts. *Journal of Conflict Resolution* 24:285–88.

Weiss, Herbert. 1966. Combat models and historical data: The U.S. Civil War. *Operations Research* 14.

Willard, D. 1962. *Lanchester as force in history: An analysis of land battles of the years 1618–1905.* RAC-TP-74. Bethesda, MD: Research Analysis Corporation.

Wolpin, Miles. 1986. *Militarization, internal repression, and social welfare in the Third World.* New York: St. Martin's Press.

Wolpin, Miles. 1991. *America insecure: arms transfers, global interventionism, and the erosion of national security.* London: McFarland and Company.

Wright, Quincy. 1942. *A study of war.* 2 vols. Chicago: University of Chicago Press.

Wulf, Herbert. 1979. Dependent militarism in the periphery and possible alternative concepts. In *Arms Transfers in the Modern World,* edited by Stephanie Neuman and Robert E. Harkavy. New York: Praeger.

Zagare, Frank. 1987. *The dynamics of deterrence.* Chicago: University of Chicago.

Zagare, Frank. 1990. Rationality and deterrence. *World Politics* 62:238–60.

Zanoni, Michael. 1989. The shopping list: What's available. In *The International Legal and Illegal Trafficking in Arms,* edited by Peter Unsinger and Harry More. Springfield, IL: Charles Thomas.

Index